"Learn Technology,
Learn The Computer,
Learn Your Desktop"

(THE L3)

"Learn Technology, Learn The Computer, Learn Your Desktop"

(THE L3)

Enter a Tech Savvy World
Still in Exploration
by Great Enthusiastics.
Be A Part.

Joko Austin Bryant

To order additional copies of this book, contact:
Xlibris Corporation
1-888-795-4274
www.Xlibris.com
Orders@Xlibris.com
116380

"Learn Technology, Learn The Computer, Learn Your Desktop"

(THE L3)

Enter a Tech Savvy World
Still in Exploration
by Great Enthusiastics.
Be A Part.

Joko Austin Bryant

Introduction

This book is being written with two major thoughts in mind. It can be used as a textbook whereby students in the classrooms up to secondary education, or even college education, can enrich their minds and build a solid foundation with a basic education in computer technology and other related technologies. This is true because the materials presented herein are not limited to the computer, how it evolved, who the key players have been, how its functions have transformed from inception till today, but also how it interrelates to technologies that have come along, whether or not they stemmed directly from the original computer. The student will craft his/her mind with knowledge that is crucial to success by reading and exploring new discoveries that can be applied to the real world. The student will discover that the computer provides an enhanced sphere of learning, working, interacting, and staying on top of things, because the computer can be separated from hardly anything in the classroom or workplace in this new era, and this fact continues to hold at an astonishingly increasing rate.

This book is also being written with the workplace in mind. Government agencies and offices, corporations, small businesses everywhere the world over, should have this specific goal toward success—increase productivity. After addressing so much about the computer and how it came into being, the book goes further to delve into the types of computers we use today and the latest technologies in relation to the computer also in use. Industries cannot function properly without efficient and effective secretaries and office workers. And to possess one or both of these qualities, the employee must adapt to new steps at one point and devise strategies that improve the work process.

Upon the provision of a general overview of the computer, the latter chapters turn to the specifics of the personal computers (PCs) and the Microsoft operating systems and user environments. In the Microsoft Windows computing systems, many functions can be performed using the standard methods with the keyboard and mouse, and again in the shortcut methods with key combinations that speed up the task at hand, and make the computing experience more rewarding and exciting.

Computer technology has grown significantly over the years since inception, made dramatic changes, and taken various forms, types, and shapes. Computer types have changed from the mainframe of yesterday to the personal computers, tablets, and handheld devices today. No matter how one sees it, the average smart phone and personal digital assistant (PDA) have proven themselves to be acceptable handheld computer types, considering the many functions they can execute aside from making a simple phone call.

Again, the purpose of this book is to provide the reader/learner with a basic overview of the knowledge of not only the desktop aspect of computing, but a broad perspective of its hardware architecture and software types and their functions. In such context, the term "academic" refers to the learning of technology through reading, feeling, experiencing, and exploring the different aspects of a piece of technology. That goes beyond the typical street-sense knowledge of the role technology plays in our daily lives.

What Is A Computer?

A computer is an electronic device designed to accept data, perform prescribed mechanical and logical operations, and display the results of these operations. The computer is also referred to as a "processor," a name given to the most powerful chip in the computer, which performs tasks ranging from the simplest to the most complex.

A desktop computer

A laptop computer

The History of Computers

While on the learning curve to discover some good things you can do today through the use of the computer and related technologies, it should help considerably to learn a little about whence, how far, and from whom all of it evolved. This book guarantees that the full details of all sought information that address those subjects will not be found here, but obviously, the table below will provide you with a reliable foundation. It will be worthwhile to the reader to utilize every piece of information of interest herein for further research for the purpose of attaining a broader knowledge of the history of computers.

The table provides a statistical overview of the entire age of the computer—its birth, life, growth, and many transitions, which now shape a significant aspect of the livelihood of mankind worldwide. Let's begin.

The Computer-Age Table

The Designer/ Contributor	The Time Period	The Computer Age/Name	The Functions
	Mid-1800s to 1930s	**Early Mechanical Computers**	
Charles Babbage		- Difference Engine # 1 - The Analytical Engine - Difference Engine # 2	These concepts included the idea of separating storage from processing, the logical structure of computers, and the way that data and instructions are inputted and outputted.
Konrad Zuse		- Z1	Used to take the U.S. Census in 1890.
	1930s	**Electro-mechanical Computers**	
Gaspard-Gustave Coriolis		- The Differential Analyzer	Designed to solve differential equations by integration.
George Stibitz		- Model K - Complex Number Calculator	Early digital calculating machines.
Konrad Zuse		- Z3	The Z3 used floating-point numbers in computations and was the first program-controlled digital computer.
Alan Turing		- The bombe	Used in WWII to decrypt German codes.
	1940s	**Electronic Computers**	

Frederic C. Williams Tom Kilburn George Tootill		Manchester Small-Scale Experimental Machine (SSEM) nicknamed "The Baby"	The world's first stored-program computer
John Mauchly John Presper Eckert		- ENIAC	The first general-purpose electric computer capable of being programmed to solve a full range of computing problems. They used decimal numeric systems.
John V. Atanasoff Clifford Berry		- Atanasoff-Berry Computer (ABC)	The first electronic digital computing device.
Howard H. Aiken IBM		- Harvard Mark 1	An electromechanical computer.
Tommy Flowers		- Colossus Mark 2	The world's first electronic digital programmable computer. Developed in WWII to decrypt secret German codes. It used vacuum tubes and paper tape.
	1950s	**The First Commercial Computers**	
J. Lyons and Company		- LEO (Lyons Electronic Office)	The first computer to run a regular routine office job (1951).
J. Presper Eckert John Mauchly		- UNIVAC	The first commercial computer developed in the U.S., with its first unit delivered to the U.S. Census Bureau. The first mass-produced computer.

IBM		- IBM 701	The first mainframe computer produced by IBM
	Mid-1950s	**Transistor Computers**	A. Era of replacement of vacuum tubes and significantly smaller computers. B. Transistors also led to developments in computer peripherals.
Intel (the 4004) Texas Instrument (the TMS 1000) Garrett AiResearch (the Central Air Data Computer or CDC)	1960s	**The Microchip (integrated circuit) and the Microprocessor**	Era of the production of minicomputers and microcomputers, which were small and inexpensive enough for small businesses and even individuals to own.
	1970s	**Personal Computers**	The first personal computers were built in the early 1970s.
		- Altair 8800	First popular computer using a single-chip microprocessor.
		The Trinity - The Commodore PET - The Apple II - Tandy Corporation's TRS-80	These early PCs had between 4kB and 48kB of RAM. The Apple II was the only one with a full-color, graphics-capable display and eventually became the best seller among the trinity, with more than 4 million units sold.

	1980s and 1990s	**The Early Notebooks and Laptops**	
		- Osborne 1	The first commercially successful portable microcomputer. It had a tiny 5-inch display screen and weighed 23.5 pounds. By today's standards of portable computers, the Osborne 1 would not make the list.
Manuel Fernandez		- Gavilan SC	Introduced in 1983, the Gavilan was an early laptop and was the first portable computer to be marked as "laptop."
	2000s	**The Rise of Mobile Computing**	
	Late 2000s	**Netbooks**	The first mass-produced netbook was the Asus Eee PC 700, released in 2007. They were originally released in Asia but were released in the U.S. not long afterward.

Data imported into this table is based largely on reference to the Web site article "The History of Computers in a Nutshell" from http://sixrevisions.com/resources/the-history-of-computers-in-a-nutshell.

Computer Types

Supercomputer and Mainframes

Supercomputers - This is a broad term used for one of the world's fastest and most expensive computers available today. They are built to perform complex operations. The supercomputers are designed to perform tasks such as run applications that involve immense amounts of calculations. They are also used for generating and performing tasks such as weather forecast, nuclear energy research, analysis of geological data, etc.

Mainframes - In the world of computing today, mainframes are very large, powerful, and expensive data processing systems capable of supporting hundreds, or even thousands, of users simultaneously. This type of computer is mainly employed in large organizations to execute complex tasks. Mainframes are primarily based on the UNIX, and partly the Linux operating systems, which originated as room-sized machines, housed in very large firms. Though still very capable of executing huge tasks, mainframes, today, have shrunk in size considerably and are generally deployed to perform tasks nearly similar to smaller powerful networked computers.

Minicomputers - This is a term rarely used in today's computing age that refers to an intermediate (midsize) computer which lies between a microcomputer and a mainframe. Minicomputers are multiprocessing stand-alone computer systems that served small to large business levels and provided service at any time for up to two hundred concurrent users during their age. The size and processing capability of this computer system would place it today in the category of a server.

Servers - A server or server computer is a computer system built (with the required hardware and software) as a host computer by serving the needs of other computers and users that request resources, and as a client computer by making requests from other computer sources linked to other networks or the Internet. In a hardware context, a server is a physical computer that runs complex tasks by solving problems and answering the needs of other computers or computer systems in a client/ server environment. In a software context, on the other hand, a server is the program running on a computer capable of performing the tasks required in a client/server setting.

Personal Computers - A personal computer (PC) is an individual-orientated computer system with its size, price, and capabilities suitable for general-purpose use and which is intended to be operated directly by an end user with no intervening computer operator. This computer system is comprised of a system unit (referred to by some users simply as the CPU), a display unit (monitor), a keyboard, and a mouse. The personal computer allows users to perform noncomplex functions such as document and spreadsheet creation and tracking, checking e-mails, surfing the Web, and the like.

Workstation - A high-end personal computer designed for engineering applications (CAD/CAM) and performing technical or scientific applications. Workstations are set to run multiuser operating systems and designed to be part of a local area network. They are intended for a single user at a time. Workstations require a moderate amount of computing power and are generally used for scientific and graphic-intensive tasks such as computer-aided design, architecture modeling, image processing, drafting and modeling, etc.

Desktop computer - A desktop computer is a computer designed to fit suitably on a traditional desktop along with its attached peripherals (accessories). Though built in a variety of shapes and styles, the desktop computer was originally intended to be designed in a horizontal/rectangular shape and style to sit flat underneath the display screen.

Mobile Computers

Laptop computer - A laptop computer, or simply laptop, and also referred to as notebook, is a portable type of personal computer. A laptop computer has its display screen, mouse, and keyboard attached as a single unit. The laptop is built to conveniently sit on the user's lap while out and about and not at a particular computer desk. It is operated by a battery pack which, when fully charged, provides power for up to 2 hours of computing. The laptop is light in weight and may weigh less than 6 pounds.

Meanwhile, a laptop is featured with various I/O ports that allow it the capacity and features of a full-blown desktop computer. For example, a video graphic array (VGA) port allows a connection to an external display screen, and its universal serial bus (USB) ports will allow a connection to an external keyboard or mouse. As such, you end up with a system unit, display unit, keyboard, and mouse just as the traditional desktop computer.

But have you ever thought about what's up with the different sizes of laptop computers? You're not alone. In my journey through the pursuit of knowledge on computers, I had similar thoughts and questions. Having had my own questions answered, I'd like to take the time and pass that knowledge on to you. The laptop computer type is divided into three different categories, and here they are below:

- ➢ Desktop replacements - These are larger, heavier, and wider (15-to 17-inch) screen laptops. They also have more powerful hardware than the standard, with performance capabilities of a desktop computer.
- ➢ Standard notebooks - These are the average in size and performance in laptops. They generally carry displays of 13 to 15 inches, with performance about the standard for its type.
- ➢ Netbooks - Also referred to as subnotebook computers, the netbooks are a smaller version of a typical notebook computer. They look identical to the notebooks, except that they have smaller display screens and keyboards. The netbooks have displays smaller than 13 inches and fewer features than the standard notebooks.

The netbook has only recently added some components such as the optical drive to its features that were omitted during its evolution in 2007. To improvise for the missing optical drive still on some editions, which is a useful computer component, the user is left with the choice to purchase a removable USB-driven CD/DVD-ROM drive for media and storage, to be attached as needed.

Tablet PC - A tablet PC is a type of notebook or mobile computer that provides operability through touch screen by a fingertip, a pen, or a tiny pointed device called a stylus. This crossover computer offers some functionalities of a computer and those of a phone in some cases.

Ultramobile PC (UMPC) - The ultramobile is a smaller version of the tablet PC just as the netbook is to the notebook.

Pocket PC - A pocket PC is a handheld-sized computer, termed a PDA, that typically runs on the Microsoft Windows Mobile operating system. It can function as a cellular phone, a fax sender, and a personal organizer. It is designed to enable companies to provide their employees, partners, and customers with easy access to corporate data. Most PDAs use a stylus rather than a keyboard for data input and are enabled with a voice recognition feature.

Terminologies/Definitions

Hardware: Hardware is the mechanical equipment that is used for conducting an activity. Hardware can be defined in a collective term when describing a computer, as its physical components. Hardware may also be defined as the mechanical, magnetic, electronic, and electrical devices that make up a computer system, such as the motherboard, disk drive, CPU, and other physical components within the computer.

There are many hardware components that make up a computer system. This book will begin to discuss only a few of the computer hardware components considered essential to the reader's familiarity. Basic knowledge of these components will help you build a general overview of what the computer stands for. Now let's begin our discussion.

The computer case: The case of the computer is the housing or shell within and around which all other components reside. It is defined as the primary housing for the hardware components that make up a computer. As a technical definition, this housing is also called the chassis, pronounced *"chas-ee."* You can also call this housing the *computer cabinet*.

Inside the computer chassis are hardware components such as a motherboard, technically known as system board, to which nearly all the remaining components are connected.

The system board, also referred to as the motherboard or main board: The system board is the main circuit board in the computer that directly or indirectly connects and provides power to all circuitry and hardware components that make up the computer. The system board receives its power directly from a source through the power supply unit, which it in turn distributes to multiple components through various connectors. This large piece of silicon is what integrates all the components within the computer.

Main memory: Also referred to as RAM (random-access memory), the computer memory is a chip that internally stores data in execution by the processor. It is the temporary storage area of data that will be immediately needed by the CPU for processing. The term "memory" is usually used to represent physical memory which is the actual chip. The term "virtual memory" is the capability of the computer to expand physical memory onto the hard drive.

Types of Computer Memory

1. *Computer RAM*

   The computer RAM is the best-known form of memory your computer uses. Every file or application opened is placed in RAM. Any information the computer needs or uses becomes part of a continuous cycle where the CPU requests data from RAM, processes it, and then writes new data back to RAM. This can happen millions of times in a second. However, this is usually just for temporary file storage, so unless the data is saved somewhere, it is deleted when the files or applications are closed.

2. *Hard Drive*
    A hard drive is a form of computer memory that allows you to permanently store data. This is where all of your permanent files and programs are stored. On computers running with Microsoft Windows, the hard drive is often called C drive. The size of a hard drive is typically measured in gigabytes.

3. *Virtual Memory*
    Virtual memory typically comes into play when applications are too large for the RAM to handle. The operating system uses the hard drive to temporarily store information and take it back when needed. This is normally a lot slower than actual RAM and can possibly degrade performance if used to heavily.

4. *Cache Memory*
    Cache memory is used in between the CPU and the RAM and holds the most frequently used data or instructions to be processed. There are three different grades of cache. Some systems will only have level 1 and level 2. More advanced systems will include the level 3.

    1. Level 1 (L1) - This is the primary and is on or very close to the processor. This is used for the most frequently used data and instructions.
    2. Level 2 (L2) - This is second closest to the CPU and is more common to be on the motherboard. Depending on your motherboard, it might be able to be updated. This is used for the most frequently used data and instructions.
    3. Level 3 (L3) - This is the most advanced cache and will speed up the memory even further. This is used for the most frequently used data and instructions.

Types of Computer Memory. Retrieved March 18, 2012, from http://www.computerknowledgeforyou.com/computercomponents/ types_computer_memory.html.

Processor (CPU) - The central processing unit, abbreviated CPU, is

The CPU underside view

The CPU top view

commonly referred to as the brain (others may call it the engine) of the computer. It is the most important element of the computer system because of the functions it executes. The CPU is the unit of the computer responsible for performing the instructions of a computer program. This includes the basic arithmetical, logical, and input/output operations of the system. The speed and power of the computer mostly rests with the CPU. The physical chip which houses and represents the CPU is called a microprocessor. This is the modern implementation of the CPU for personal computers since the 1970s.

An internal component of the computer, modern CPUs are small in size, square in shape, and bear multiple metallic pins on the underside, systematically designed to fit directly into corresponding pin holes arranged in a CPU socket on the motherboard.

CPU Types: The world's two major processor manufacturers and suppliers for computers, Intel and AMD, have released several versions of desktop and laptop computer processors. At the time of this writing, Intel is the brand name leader of the two. Intel has released from its most basic yet, the Intel Celeron, to the most versatile, the Intel Core i7 and i7 Mobile. On the AMD side, the Athion Neo and Neo fall at the bottom, and the Phenom II X6 is at the very top in comparison.

For the purpose of this book, we are directing the reader's attention to the table below, whereby three specification levels are outlined for CPUs by the world's two largest manufacturers of processors. The table, accompanied by all of its contents below, was retrieved from the University of Wisconsin Department of Information Technology Web site, as follows:

Intel & AMD Processors
Breaking down the *What, Why* and *Which*

1. What's the big deal about choosing a processor?

The processor is the "engine" of a computer. It is the most important component in determining how fast or "snappy" the system will operate across applications both now and in the near future. Like the engine of an automobile, a processor can be fast, slow, power hungry, or power efficient subject to the kind of work the computer is being considered for. It is important to round out what kind of things you will be doing on the system to best select a computer with a CPU most suitable to your needs.

Unlike other components of a notebook computer, the CPU is—with rare exception—a fixed component. This is in contrast to RAM and hard disk storage, which can typically be upgraded. Therefore, another consideration is the fact that (important as the CPU is) the CPU you choose will be the same throughout the life of the system. This implies that as programs become more sophisticated, the computer's ability to handle such programs will be directly affected by the decision made at purchase all that time ago. This choice may mean the difference between a system that is useful for another year or two versus one that isn't much sooner. A final consideration in choosing a CPU is the suggested or minimum requirements of either the programs that is planning on being run, or academic department recommendations as a guide as to the relative kind of performance required for a particular field of study.

II. Product Line Comparisons Hierarchy

Currently, the two largest manufacturers of CPUs in the world are Intel and AMD. The following provides a short profile of the companies and the current state of their products.

Intel

The current performance and market leader at the time of this writing is Intel. Intel is currently the sole supplier of processors for all recent Apple computers (Macbook, Macbook Pro, Mini, iMac, etc.) and are found in virtually all major computer manufacturers' product lineups. Intel's most current crop of CPUs are the Core iX-series processors which include the i3, i5 and i7. As of January 2011, these series of processors entered their second generation (codenamed "Sandy Bridge" where the first generation was codenamed "Nehalem," with their differences explained under the special features section).

AMD

AMD is the second largest supplier of processors for personal computers. Many of their products are found in both high-performance and budget-oriented notebooks as well as low-cost, enthusiast-oriented desktop builds. The Phenom II and Fusion platforms comprise AMD's most popular and mainstream offerings at the time of this writing. The Fusion processors are a line of AMD processors that contain a graphics procession unit integrated with the processor. The Fusion line offers similar processors from the energy-efficient Atom processor up to the powerhouse Core iX series.

Beneath, we provide a chart which compares the relative performance between competing product lines within Intel's and AMD's offerings. These are organized by the following three classes: high-end, midrange, and economy. It is important to note that though this comparison offers a reference of relative performance within each brand, it does not necessarily indicate absolute rankings between competing Intel and AMD products (for instance, the Core i7 is in the same row and category as the Phenom II series but offers superior general performance). Further, the Core iX Mobile series only indicates relative performance for notebook platforms—that is, it is generally not useful to compare them to desktop processors such as the Intel Core i7 or the Phenom II series.

High-End Processors
Intensive Statistical Analysis, Professional Video/Audio Creation, Advanced 3D Graphics

Intel Core i7	**Intel Core i7 Mobile**	**AMD Phenom II X6**
As Intel's flagship processor, the i7 is a 64-bit processor offering either two, four, or six cores of the highest levels of general performance available. The i7 combines Hyper Threading and Turbo Boost technologies for the most demanding and advanced of applications.	Intel's Core i7 Mobile features unparalleled performance on notebooks, incorporating significant power savings while implementing the same features as the nonmobile i7, Hyper Threading and Turbo Boost. The i7 Mobile is available on notebooks with two or four cores; currently the four-core version offers higher performance in some respects, but heat and battery life are concerns.	AMD's Phenom II X6 represents the industry's first consumer class six-core processor. The X6 offers the highest levels of performance ideal for the most intensive of tasks—bolstered by AMD's new Turbo Core technology, the X6 is able to optimize performance in a variety of situations.
Intel Core i5	**Intel Core i5 Mobile**	**AMD Phenom II X4**
Based upon the same architecture as the i7, the i5 is also a 64-bit processor that features two or four cores at a similar class of performance of the i7 processor at a lower cost. The i5 features Turbo Boost and Hyper Threading technology, but do not possess as much cache memory as the i7.	The Intel Core i5 Mobile, while also featuring Hyper Threading and Turbo Boost, possesses a similar but a lesser class of performance than the Core i7 Mobile, with less cache and available in notebooks only with two cores. The Core i5 Mobile is a high-performance processor with low energy requirements.	AMD's latest generation of consumer class four-core processors, the quad-core Phenom II X4 chips are designed to deliver performance ideal for all kinds of multimedia as well as in the most demanding of applications such as virtualization.

Intel Core i3	Intel Core i3 Mobile	AMD Phenom II X3 and X2
Derived from the same architecture as the higher-end i5 and i7, the i3 is available strictly as a dual-core processor. Though Hyper Threading is available, it does not feature Turbo Boost. The Core i3 processor presents higher levels of performance than the Core 2 at a smaller cost.	The Intel Core i3 Mobile descends similarly from the i3, presenting a fast, 64-bit computing experience with the intelligent architecture of the i5 Mobile and i7 Mobile. The i3 Mobile features two cores and Hyper Threading, but does not include Turbo Boost technology.	AMD's Phenom X3 and X2 processors boast three and two cores that offer excellent performance value; great for all around usage on a small budget, all while utilizing AMD's latest architecture technology seen in the Phenom II X4 series.

Intel Core 2 Quad

The Core 2 Quad features four processing cores to optimize gaming, video, and image processing. Built on the same architecture as the Core 2 Duo, this processor excels on multitasking with performance-hungry applications.

Intel Core 2 Extreme

Available in both two- and four-core versions, distinguishing features of the Extreme series include higher bus speeds than the nonextreme versions, and an unlocked clock multiplier for further customization of your computing performance.

Midrange Processors
Speed & Multitasking, Adobe Creation Suite, All-around Use, Basic 3D Graphics

Intel Core 2 Duo	**AMD Phenom I X3 & Phenom I X4**
Contains two processing cores to optimize gaming, video, and image processing. Laptops with this chip tend to be thinner and more energy-efficient.	AMD's first generation of consumer class processors featuring quad- and triple-core performance found in desktop builds. Features 64-bit computing performance, as well as AMD's HyperTransport bus technology.
Intel Pentium Dual Core	**AMD Turion II Ultra / AMD Turion II**
Dual-core processor based on the Core microarchitecture. A class beneath the Core 2 Duo and Core Duo of Intel's processor offerings, the Pentium Dual Core is available in current desktops and laptops.	The Turion II and Turion II Ultra are AMD's mainstream mobile processor platform; they provide excellent all-around performance for multimedia such as high definition video. As these are often paired with AMD/ATI graphics, budget configurations containing these processors are also sufficient for basic 3D graphics and gaming.

Intel Core Duo / Intel Core Solo	**AMD Athlon II X2**
The Intel Core Duo and Core Solo are dual- and single-core processors based on the Core microarchitecture. The Core Duo and Core Solo offer modest performance for office and limited multimedia oriented tasks.	The AMD Athlon II X2 is a two-core desktop processor that is 80% faster than its single core counterpart. Great for multitasking and multimedia consumption on a budget.

Economy Processors
Internet Browsing, E-mail, Microsoft Office, Simple Graphics and Games

Intel Centrino/Centrino Duo	**AMD Sempron**
A mobile-oriented processor based upon Pentium M or Core Duo architectures, the Centrino also integrates wireless networking technology, allowing for smaller-sized laptops. Offers slight performance boost over simply choosing a core duo and Dell wireless card (which is typically less expensive.)	The AMD Sempron is a budget-class processor seen in low-cost notebooks and desktops, and are considered a class above netbook/nettop processors such as the Intel Atom or the AMD Neo platforms.

Intel Atom	**AMD Athlon Neo / Neo X2**
Primarily found in netbooks and nettops, this processor was designed with price and power consumption in mind. As a result, it offers much less processing power than other current Intel alternatives. This processor is available in one or two cores, with the single-core option being far more prevalent.	The Athlon Neo and Neo X2 are single- and dual-core processors seen in ultramobile platforms such as netbook and nettops. They are featured with ATI integrated graphics for reasonable multimedia playback performance.

Intel Celeron

Intel's economy model processor. It is the most basic, and thus the slowest. It has less cache than other Intel processors, so even if it has the same GHz rating as another processor, it will be slower. We usually do not recommend this processor because it offers the least in terms of longevity.

University of Wisconsin DoIT Showroom, Intel & AMD Processors. Retrieved April 17, 2012, from https://kb.wisc.edu/showroom/page. php?id=4927.

Power supply - A power supply, sometimes referred to as a power supply unit (PSU), is a hardware component that supplies power to an electrical device. It receives energy from an electrical outlet and converts that energy from AC (alternating current) to DC (direct current). The computer requires direct current in order to function. The power supply has an input socket to which a power cord connects from the main power source. Its corresponding output socket is connected to a power cord which makes direct contact (connection) to the power source (outlet).

Power supply   Power supply
Side view   Top view

Some power supplies do quite a good job at regulating the electrical voltage to avoid over usage or under usage that may cause damage to the computer.

Hard Drive - The hard drive is the storage mechanism that provides prominent storage of data (information) for the computer. The hard disk drive (HDD) is the housing unit of the hard disk inside the drive, which provides the actual reading and

Hard disk drive.
Top view   Hard disk drive
                    Underside view

writing of data. Even though called "hard drive" for short, the device is called "hard disk drive" describing the drive that holds the hard disk.

Peripherals

A computer peripheral is an external device connected to the host computer to expand its capabilities, but not necessarily a part of the computer. The peripheral provides input/output (I/O) support for the computer, and is called an input/output device.

Input device - An input device is any device or tool used in computing to feed data or instruction into the computer for display and outputting or transmission. An input device can be a component or peripheral device, such as a mouse, keyboard, scanner, barcode reader, joystick, or stylus.

Output device - An output device is a component used in computing to display or transmit processed data or information.

Below are some input/output devices:

Monitor (Display Unit) - In computers, a monitor is a separate unit of hardware designed to display processed data and information on the computer. It is a separate physical unit from the computer, and is attached via a display port, such as a VGA or DVI. Technically, the terms *monitor* and *display* are used interchangeably even though a display may be an attached area of a device. Notebook computers don't have monitors, though they may use one by having it connected via VGA port which most computers come with. They (notebooks) have attached display screens. A monitor (display) is considered an output device to the computer.

Keyboard - A computer keyboard is an input device with typewriter-type keys that enable the user to enter data into the computer. The computer keyboards are similar in key formulation to electronic typewriter keyboards, but contain additional keys.

Numeric keypad - A numeric keypad is a separate set of keys that may be found on a computer keyboard, a calculator, an adding machine, or any other electronic device. The keypad on the keyboard is located on the far right hand side, consisting of numbers 0 to 9, the Enter key, the Num Lock key, the forward slash (/), the asterisk (*), the plus sign (+), the period (.) and the Del key. The numbers on the keypad share functions with other characters at the execution of certain special key combinations. The keypad is an input device, because it feeds data or information into the system.

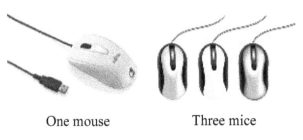

One mouse   Three mice

Mouse - A mouse (plural: mice) is a pointing device that physically is held under the user's hands, with one or more buttons. In addition to the principal clicking buttons, the mouse may feature a scroll wheel, which may also act as a third button. The mouse is a computer input device.

Display Types

In order to foster a brilliant discussion on the types of display used with computer/electronic devices, I picked an HP LaserJet printer and captured some status pictures from it. Below, you will see the pictures used to exhibit the different statuses shown by LEDs and the LCD panel on the device.

LED - A light-emitting diode (LED) is a semiconductor diode that emits visible light when an electric current passes through it. LEDs are mostly used in flashlights, and can be used in flat-panel computer displays and for displaying readings on electronic equipment such as digital watches, calculators, etc. It is typical for LEDs to display flashing lights as well as solid lights to indicate the status of activity of the device.

Exhibit A   Exhibit B   Exhibit C

Notice the exhibits shown above. In A, the LED status light displays a solid "Ready" to indicate that the printer is ready to accept data for processing (printing). In B, the LED status light displays a blinking (flashing) "Data" light to indicate that the printer is accepting data and ready to process (print). In C, the LED status lights display a solid "Data" and flashing amber "Attention" to indicate that the printer is now receiving data to process but there is a need for attention to some aspect of the printer functions.

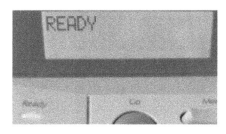

LCD - A liquid crystal display (LCD) is a very thin display type used in many portable computers and flat-panel monitors and aircraft cockpit displays. They are common in electronic devices such as clocks, watches, video players, calculators, and telephones, just to name a few. LCDs have replaced the legacy cathode ray tube (CRT) in many cases.

From pictures captured from the HP LaserJet printer mentioned earlier, the LCD panel above displays a "Ready" also as shown on the solid green LED lighting below the panel, indicating that the device is ready to accept data for processing (printing).

CRT - A cathode ray tube (CRT) is the technology used in older television and computer display screens. The display unit built with the CRT technology is referred to as a CRT display or terminal. The CRT is the older, larger, bulkier, and heavier of the two basic types of computer monitors ever used. Around the time of this book, CRT devices including computer monitors and televisions have rapidly become extinct and replaced by LCD technology, though there are still a few around in some areas.

You have been provided with an overview of many of the key hardware components that make up the computer, including some of the key hardware components inside the computer and some input/output peripheral devices. We will now discuss the ports on the computer, from the legacy types to the most up-to-date, around the time of publication. Note that as technologies change and newer ideas and devices are employed, some of the older ones will have to be replaced. Newer computers are designed by manufacturers to feature ports suitable for the latest technologies of the era. We will now continue to discuss some of the ports found on the computer that provide new as well as legacy connection interfaces.

Computer Ports

A computer port is an interface or point of connection on the computer to a peripheral or another device. A port can be seen as either physical or virtual, depending on the context of use of the term. A physical port allows the computer to connect to another device physically. A virtual port in networking is a virtual point of contact, usually through the Internet via TCP/IP, by which networked computers recognize each other and communicate.

Let's begin.

IEEE 1394 Interface (Say the term in this technology as "I triple E")

FireWire (IEEE1394)

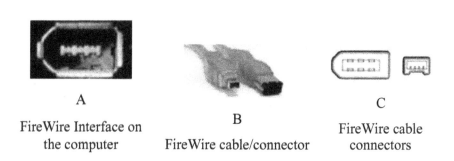

A
FireWire Interface on
the computer

B
FireWire cable/connector

C
FireWire cable
connectors

The IEEE 1394 (FireWire) technology was developed by Apple Inc. in the late 1980s and early 1990s. It is a high-speed communications data transfer technology. FireWire is a high-definition audiovisual (AV) communications technology.

There are two primary versions of the FireWire interface standard, and they are (1) FireWire 400 (IEEE 1394a) and (2) FireWire 800 (IEEE 1394b). FireWire 400 supports data transfer at the rate of up to four hundred megabits per second (reading 400 Mbps or 400Mb/s). This FireWire interface has a 6-pin connector. FireWire 800, on the other hand, supports data transfer at the rate of up to 800 Mbps and has a 9-pin connector.

There has been quite a few Internet blogs (Weblogs) that discuss which is a better, faster, and most adaptable communications connector/cable type between the FireWire and the USB. In my mind, the most practical tech-savvy activists have been the best judges.

Ethernet - (Be careful how you say this technology term (E-ther-net) when in the presence of a technical novice. It may be understood as "Internet," from which it has a huge difference.

Exhibit A-1	Exhibit A-2	Exhibit B-1	Exhibit B-2
Ethernet RJ-45 port	Ethernet RJ-45 cable/connector	Telephone jack RJ-11	Telephone RJ 11 Cable/connector

"Ethernet" is a terminology for a type of local area networks (LANs) developed by engineer Robert Metcalfe in the 1980s. On a TCP/IP network, Ethernet networking runs on physical and data link layers. When first developed, Ethernet supported a data transfer rate of up to ten megabits per second (Mbps), a technology also termed 10BaseT. Later, the development took another stage when Ethernet networking reached the capability of data rate of 100 mbps, called Fast Ethernet, and also termed 100BaseT Ethernet. Today, Gigabit Ethernet (GbE or 1GigE) performs a high-speed data transfer of one thousand megabits per second, equivalent to one billion bits per second.

The Ethernet connection interface has some similarity to the regular phone jack, but wider, as well as its corresponding cord, but larger. See exhibits above. A difference between the two is also seen in their technical terms, where Ethernet technology is termed RJ-45, short for Registered Jack-45, while the regular phone jack is termed RJ-11. Another difference to mention between the two types of Internet connections is that RJ-11 will use a telephone line or modem to connect to the Internet, as opposed to RJ-45 which connects directly to the local area network (LAN) through the Ethernet port. Also, the port of the RJ-45 interface or a typical RJ-45 connector has eight pins. Correspondingly, the RJ-45 cable has eight wires that connect to the cable interface.

To sum it up, Ethernet networking is used to connect your computer to the Internet at and about the speeds indicated above and communicate with other computers and network devices attached to other networks and the Internet elsewhere.

PCMCIA

PCMCIA / Cardbus (WiFi, Networking and expansion cards)

The PCMCIA/CardBus port on the laptop computer provides for Internet connectivity using the PCMCIA card also known as PC card. The card was defined and developed by the Personal Computer Memory Card International Association (PCMCIA). PCMCIA began as the name of the association that developed the standard, which was then named after the association itself.

The PCMCIA slots on a computer come in three sizes to accommodate three sizes of PC cards. There are also three types of PCMCIA cards. They are all rectangular and measure 8.56 by 5.4 cm, but have different widths:

- Type I: up to 3.3 mm thick, mainly used to add additional ROM or RAM.
- Type II: up to 5.5 mm thick, typically used for fax/modem cards.
- Type III: up to 10.5 mm thick, often used to attach portable disk drives.

Here are explanations of the three sizes in which the PCMCIA slots have been available on the computer: A Type I slot can hold one Type I card, a Type II slot can hold one Type II card or two Type I cards, and a Type III slot can hold one Type III card or one Type I and one Type II card. PC cards can be removed or inserted on a "hot swap," which means you don't have to turn your computer off to exchange them and you don't have to restart your computer for the system to recognize them.

This computer port is still popularly found on many newer notebook computers today in spite of several newer technologies that can support the features and functions it is featured with.

Parallel/Serial/Game Ports

Standard parallel interface   Serial interface (RS 232)   Game port

Parallel Port

IEEE 1284 (Standard) Parallel Interface (25-pin)

Exhibit A   Exhibit B   Exhibit C

IEEE 1284 parallel   IEEE1284 parallel
printer cable interface   printer cable interface
on the computer   on the computer
(36-pin female)

The parallel port is a parallel communication physical interface found on the back of older computer products such as PCs, printers and scanners, and on older test equipment. This 25-pin connector (type DB-25) technology is defined by the IEEE 1284 standard, and is commonly known as a printer port. The parallel port is also technically known as a Centronics Port, after the company that first developed the standard for parallel communication between a computer and printer.

Today, a newer and more advanced technology has upgraded the interface and is named the EPP (Enhanced Parallel Port) or ECP (Extended Capabilities Port), which supports the same connectors as the original Centronics interface. Both interfaces support bidirectional communication allowing the transmission and reception of data bits at the same time. They also support much faster data transfers.

As is the case with other legacy connection types (ports), the Centronics interface is adaptable to the new USB technology to allow connection to newer computers and external devices.

Serial Port

RS-232 Standard (9/25-pin) Serial

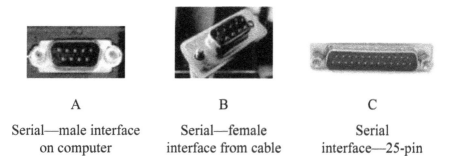

A	B	C
Serial—male interface on computer	Serial—female interface from cable	Serial interface—25-pin

The serial connection/data throughput involves the transmission of data one bit at a time over a single communication line. This type of connection is used between older computers and other peripherals and communication devices, such as mice, older printers, modems, and gaming controllers. This can also be found on long-distance-communication devices, such as credit card machines.

The serial interface is referred to technically as an RS-323 interface, and comes in two types, as can be seen in the exhibits above. It comes in a 9-pin connection referred to as DB9, and the 25-pin connection, DB25. In the picture below, the serial adapter has the 9-pin interface at one end and a USB interface at the other.

USB to serial interface

Game Port

Computer game port is the input/output (I/O) connector that is used to attach controllers such as a joystick or other gaming devices to a computer.

PS/2 Connector

A

P/S2 keyboard
(purple) and mouse
(green) interface
on the computer

B

PS/2 interface from
the keyboard or mouse
cable

C

6-pin PS/2 interface
on the computer
(female) or the cable
(male).

The P/S 2 connector takes its name from IBM's Personal System/2 computer system developed in 1987. In the years following its debut, the PS/2 port generally referred to as the mouse port, became standard for mouse and keyboard connections for all IBM-compatible computers. Today, at the time of this book, the PS/2 connector on computers is nearly extinct. Any computer with this type of I/O port is generally considered as an older PC.

As seen in exhibit A, the two PS/2 ports on the back of the older computer are circular in shape with six pins (female) for interfacing with a male connector attached to the mouse or keyboard device. As a principle color-coding scheme on the computer, the keyboard port is purple and mouse port green.

What you need to know about the position of both ports (mouse and keyboard) at the back of the computer:

When placed in an up and down position, whereby the computer power supply is located above or below, the keyboard is much likely to be found toward the end of the power supply. When placed horizontally, on the other hand, the keyboard port is much likely to be found toward the outside of the chassis and the mouse port toward the inside.

The case scenario comes handy and is more important and useful to computer users who may be inclined to using their imagination to connect any of the two peripheral devices mentioned without direct sight to the back of the computer.

Universal Serial Bus (USB)

USB 1.1, 2.0, and 3.0

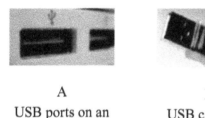

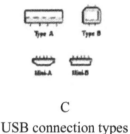

A	B	C
USB ports on an electronic device	USB cable and connection interface	USB connection types

Universal Serial Bus, abbreviated USB - This is a type of connection that has generally replaced several legacy types of connections used. This connection type is used for communication and power supply between computers and different types of devices. USB refers to the types of cables, connectors, and ports that connect computers to other types of devices.

The word "universal" in USB could be perceived from the fact that many older connection types are convertible to the USB type by interfacing the two different types. This is accomplished in many cases by the use of an adapter. The serial, parallel, and game connectors found on older computers are a few that now become converted very easily to a newer device by the use of an adapter. Generally, the adapter bears the USB interface at one end, and the legacy interface at the other to complete the connection.

The USB-to-a-legacy-device connection is just one use of the type of universal communication. The connection type has become the standard for connection to many types of devices of the same connection type, such as hard drives, cameras, phones, camcorders, mice, modems, and keyboards to a computer.

Like the IEEE, the USB is an industry standard developed in the mid-1990s, and it supports plug-and-play and hot plugging. That means no driver software installation is necessary for the USB device, and you can connect the device for use while the computer is already running or the connecting device is already in use.

USB hub - A device that increases the number of ports to connect additional devices. The hub is usually attachable to the computer desk to allow for convenient connection of other devices. Because one connector needs to be plugged into the computer, the total number of ports said to be available on a hub is minus the one. For example, an eight-port hub will have one port plugged into the computer and add seven new ports.

Here are data transmission rates:

- USB 1.0: 1.5 Mb/s
- USB 1.1: 12 Mb/s
- USB 2.0: 480 Mb/s
- USB 3.0: 5.0 Gb/s

Note: Mb/s (Mbps) is brief for megabits per second (mega = 1 million). Gb/s (Gbps) is brief for gigabits per second (giga = 1 billion).

At the time of this book, the USB 3.0 is the highest on the speed and performance chart of USB technology.

Popular Computer Video Connector Types

For the purpose of this book, you will find the minimum of the universal video connection types, and only those found most popularly on a computer, notebook, or desktop. Other types such as the DVI (mini or micro) have become increasingly popular on modern computers and video cards.

Video Graphics Array (VGA)

A
VGA port on a
computer

B
VGA cable/connection
interface from display
device

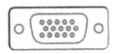

C
View of a 15-pin
male/female video
connection interface.

A. The Video Graphics Array (VGA) video interface is commonly used to connect the computer to an external display or video output system. The port on the computer is a fifteen-pin (female) connector that is interfaced with the connector provided by the video device that connects to the computer.

B. The VGA cable connection from the device is usually a cable supplied by the video device to be connected to the computer with the matching fifteen-pin (male) connector.

C. A graphical appearance of the VGA cable/connection interface:

Digital Visual Interface (DVI)

A
DVI port on the back
of a computer

B
DVI cable/ connection
interface from a
display device

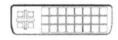

C
View of a 24-pin male/
female DVI port

DVI Interface: The Digital Display Working Group (DDWG), industry consortium, developed the Digital Visual Interface (DVI) standard to convert analog signals to digital signals to accommodate both analog and digital monitors, leaving the technology with the abbreviated name, DVI. The signal conversion is accomplished by the type of connection and display. The DVI connection converts a digital signal to an analog signal for an analog display. At the same token, no conversion takes place if the display is a digital one.

S-Video Interface

S-Video Ports for video Input/Output

S-Video - Abbreviated from "Super Video," S-Video is a technology for transmitting video signals over a cable by dividing the video information into two separate signals. S-Video input jacks (ports) are found on some laptops and electronics such as TVs, camcorders, cameras, and it provides moderate picture quality with the use of a corresponding S-Video cable. The video transmitted through this cable type is not accompanied by audio. You will need a separate cable that will produce audio.

High Definition Multimedia Interface (HDMI)

HDMI - This is a specification that provides a digital interface for audio and video transport (communication) through a single-cable solution for home theater and consumer electronics equipment. Since its introduction

A
HDMI (female)interface (on the computer/ electronics device)

B
HDMI (male) plug (usually on the cable)

in 2002, the technology has worked popularly with digital versatile disc (DVD) players, digital television (DTV) players, monitors and visual display units, AV receivers, satellite cable boxes, and other popular consumer electronics.

The HDMI technology supports high bandwidth for content delivery, and has proven to be highly standardized in support of high-definition video together with crisp, multichannel surround sound audio through its delivery capabilities with those features.

Color-coding Scheme

For an informational purpose through this book, the color-coding scheme (guide) below was retrieved from Wikipedia.org. The reader should consider this guide as a great reference tool to facilitate ease with connecting various types of input/output devices to the computer, some of which have been addressed above under computer ports.

Color	Function	Connector on PC
Mouse and keyboard		
Green	PS/2 mouse / pointing device	6-pin mini-DIN female
Purple	PS/2 keyboard	6-pin mini-DIN female
General input/output		
Black	USB	USB Type A female
White	USB 2.0	USB 2.0 Type A female
Sky blue	USB 3.0 SuperSpeed	USB 3.0 SuperSpeed Type A female
Gray	IEEE 1394 (FireWire)	6-pin FireWire 400
Burgundy	Parallel port	25-pin D female
Teal or turquoise	Serial port	9-pin D male
Video		
Blue	Analog monitor	15-pin VGA female

White	Digital monitor	DVI female
Yellow	S-Video	4-pin mini-DIN
Yellow	Composite video	RCA jack

Audio

Pink	Analog microphone audio input (mono or stereo)	3.5 mm TRS
Light blue	Analog line level audio input	3.5 mm TRS
Lime green	Analog line level audio output for the main stereo signal (front speakers or headphones)	3.5 mm TRS
Black	Analog line level audio output for the surround speakers (rear speakers)	3.5 mm TRS
Silver	Analog line level audio output for "side speakers"	3.5 mm TRS
Orange	Center speaker / Subwoofer	3.5 mm TRS
Gold	Game port / MIDI	15-pin D female

Wikipedia, Color-coding scheme for connectors and ports. Retrieved May 7, 2012, from http://en.wikipedia.org/wiki/PC_System_Design_Guide.

Sound Card Color Code

Sound card color code:	
Color	**Connector**
Lime Green	Line-out, front speakers, headphones
Pink	Microphone
Light Blue	Stereo line-in
Orange	Subwoofer and center out
Black	Rear surround speakers for 5.1 and 7.1 systems
Gray	Middle surround speakers for 7.1 systems
Gold	Midi / game port (joystick)

Computers are manufactured with the intention that they will produce sound just as they do video and other functions that are required by the system. Therefore, computers are generally built with an integrated sound functionality. The integrated sound is generally called by other terms, such as "on-board audio," "on-board sound," or "integrated audio."

The embedded sound on the computer's system board may go out or fail, or perhaps the computer user may see the need to improve the graphics on the computer than what is originally on board. This is where the expansion slots come in. Expansion slots such as PCI or AGP slots are more popularly included on computers today to facilitate the installation of sound cards and other types of expansion cards. These slots have their extended or upgraded versions such as the PCI-Express in the PCI family. A third type of expansion slot is the ISA, which is barely heard of today, around the time of this book. ISA slots are black in color, slower than the PCI slots, and about twice the length of the PCI slots.

The color-coding scheme and explanations above may serve as a guide for connecting many types of input/output devices to the audio module.

Software Definition:

Software is defined as a virtual or electronic program that is stored or processed on the computer. Software consists of lines of codes written by computer programmers that have been compiled into a computer program. In computing, the terms "software" and "hardware" are likely to be somewhat confusing in the sense that the software (electronic) programs are written or recorded onto disks (CD-ROMs, DVDs, and other media types), which may be a piece of hardware. When you purchase the program (software) on a disk, you receive the disk (a piece of hardware) on which the software is written. As technology takes newer turns and shapes, and the Internet revolutionizes various processes, software programs are being transferred to the end user via Internet browsers and other electronic media. This means the hardware transfer process is becoming a legacy process by each passing day.

Software is often divided into two categories:

- **Systems software:** Includes the operating system and all the utilities that enable the computer to function.
- **Application software:** Includes programs that do real work for users. For example, word processors, spreadsheets, and database management systems fall under the category of applications software.

Desktop Operations (Basics)

Desktop: The area of a computer screen that displays the icons and windows. This normally is the case in a graphical user interface (GUI) environment, meaning the user has direct access and interaction with the computing process on the computer screen which also displays the computer background. The computer desktop can be seen as the traditional desktop in the workplace or home office where items such as books, stationery, and other desk items are placed and organized. As a computer end user, think of your display desktop with files, folders, and other desktop items organized in the same context as all the items on your office desktop.

Icon: A symbol, picture, or graphic image by which something is represented. That something can be a computer program or file. As a computer user, you may see pictures on the computer screen or desktop that symbolizes a system program or application that has been installed into the computer. When the computer is installed or set up initially, it may place a number of widely used icons on the screen (desktop), such as the browser, and system folders. After the initial install, the computer user will, as needed, install application programs to help process specific tasks on the computer.

How to Add an Icon to a Windows Desktop

You may be tempted someday to wipe off Windows from your computer and reinstall. Upon completion of the installation, you notice that Windows' basic icons such as My Computer, My Documents, My Network Places, and Internet Explorer are missing. What do you do? This is how to add those icons to the desktop:

1. Windows XP - Right-click on the desktop and select *Properties* to access the Display properties. The display properties dialog box launches. Click the *Desktop* tab; click the *Customize Desktop* button below. Under the *General* tab, check the box of the Desktop icons item you want to add. **Note:** the procedure above was completed by using a short method of accessing the Display properties. The alternative method is to access the properties from the control panel. This is how: Click Start and select *Control Panel* from the menu. Make sure the panel is switched to classic view. Double-click the *Display* icon. Under the category view, select the *Appearance and Themes* category, and then click the Display icon. Continue from the *Desktop* tab as above.

2. Windows Vista and Windows 7 - Right-click on the desktop and select *Personalize*. Click the *Change desktop icons* link under *Tasks*. The Desktop Icon Settings dialog box appears. Check the box to the icon you desire and click *OK*.

Shortcut: In computers, a shortcut is an image that points to a particular program or data file. Though traditionally, a program shortcut is placed on the desktop for easy access of the associated program, it can be placed in any area desired on the computer. The shortcut of a program is different from its exact copy, and must be treated as such by the user.

Copy of an Item: You can create one or multiple copies of a program or data file on the computer. The copy is an exact replica of the original item, and it is noteworthy that the user should be very careful with how the items are treated. If it is a single application file and placed in different directories on the computer for future updates, the user must note which copy is being updated at any time. If one or more copies of any item are created in any one directory, the system automatically names the copies in numerical order as copies of the original item. If the copy is placed in a different directory than the original, the name stays unchanged.

The Taskbar: Developed since the Windows 95 version of the Microsoft Windows operating system, the taskbar sits at the very bottom of the screen and spans fully across. On the far left sits the "Start" menu, where that menu commands are executed. The right-hand area is known as the "system tray," which displays the system clock and icons, such as the battery meter, volume control, and network status. Also shown on the right side are installed applications such as antivirus software when up and running in the background. In the middle of the taskbar are open statuses of system and application programs. For example, if two sessions of Internet Explorer are opened, there will be two icons of the program showing. The system tray area of the taskbar may also be referred to as the notification area.

The Start Button: Introduced since the release of Windows 95, the Start button has remained a Microsoft Windows program command button, but with different looks. The Start button is a principal point for launching programs and tasks on the Windows computer. Its gray look, with the Start and Windows logo together, was featured in Windows 95, 98, ME, NT 4.0, and Windows Server 2003. The green Start is featured in Windows XP. The latest Windows Start design at the time of this book is in a circular Windows logo form, used in Windows Vista and 7.

As much as this book is not meant to address at length the functions of operation systems, we should note that the reader will gain a significant knowledge through the subjects of the basics and history of the Microsoft Windows operating systems. For the purpose of this book, the entire section below captioned "Windows Operating Systems History (Basics)" was researched and found at computerhope.com.

Windows Operating Systems History (Basics)

Operating Systems

MS-DOS - Short for Microsoft Disk Operating System, MS-DOS is a nongraphical command line operating system created for IBM-compatible computers that was first introduced by Microsoft in August 1981, and was last updated in 1994 when MS-DOS 6.22 was released. Although the MS-DOS operating system is not commonly used today, the command shell more commonly known as the **Windows command line** is still used and recommended.

Microsoft Windows CE - Microsoft Windows CE 1.0 was originally released in 1996 to compete in the Palm Device Assistant Category. Windows CE, as shown below, has many of the same features as Windows 95. In addition to the look of Windows 95, Windows CE also includes similar applications such as Pocket Excel, Pocket Word, and Pocket Power.

Computer Hope, Microsoft Windows CE. Retrieved January 22, 2012, from http://www.computerhope.com/wince.htm.

Microsoft Windows 3.1 and 3.11

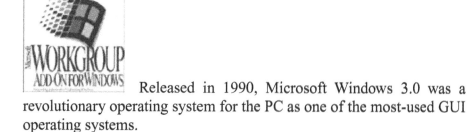 Released in 1990, Microsoft Windows 3.0 was a revolutionary operating system for the PC as one of the most-used GUI operating systems.

Later, Windows 3.1 was released in 1992 by Microsoft and was one of the first major PC GUI operating systems widely used. Windows 3.1 allowed users to utilize several features previously not available in MS-DOS. Some of these new features were the use of a mouse that allowed the user to navigate and manipulate data on the computer with one hand and easily and now did not have to memorize MS-DOS commands. In addition to the mouse, Windows now allowed the user to multitask, meaning the user could now run multiple applications at once without having to close out of each program before running another. Windows, along with other GUI operating systems are one of the many reasons computers have become easier and more widely used.

Computer Hope, Microsoft Windows 3.1 and 3.11. Retrieved January 22, 2012, from http://www.computerhope.com/win3x.htm.

Microsoft Windows 95

The next installment of Windows from Windows 3.11, Windows 95 added major enhancements when compared to Windows 3.11.

Below are some of the new features that Windows 95 has that Windows 3.x does not:

Plug and Play - Allows hardware devices to be automatically installed into the computer with the proper software. Does not require jumpers to be played with.

32-Bit Operating System - Allows the computer to run faster and more efficiently.

Registry - Combines the power of multiple configuration files into two files, allowing the system configurations to be located easier.

Memory - Windows 95 has improved memory-handling processes compared to Windows 3.11.

Right mouse click Allows you new access and text manipulation by utilizing both buttons instead of one.

CD Player - Enhanced CD player with improved usability and AutoPlay feature.

Computer Hope, Microsoft Windows 95. Retrieved January 22, 2012, from http://www.computerhope.com/win95.htm.

Microsoft Windows 98

Microsoft Windows 98 is the upgrade to Microsoft Windows 95. While this was not as big a release as Windows 95, Windows 98 has significant updates, fixes, and support for new peripherals. Below is a list of some of its new features:

Protection - Windows 98 includes additional protection for important files on your computer such as backing up your registry automatically.

Improved support - Improved support for new devices such as AGP, DirectX, DVD, USB, and MMX.

FAT32 - Windows 98 has the capability of converting your drive to FAT32 without losing any information.

Interface - Users of Windows 95 and NT will enjoy the same easy interface.

PnP - Improved PnP support, to detect devices even better than Windows 95.

Internet Explorer 4.0 - Includes Internet Explorer 4.0

Customizable Taskbar - Windows adds many nice new features to the taskbar that 95 and NT do not have.

Includes Plus! - Includes features only found in Microsoft Plus! free.

Active Desktop - Includes Active Desktop that allows users to customize their desktop with the look of the Internet.

Windows 98 SE

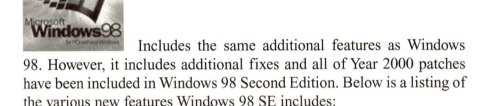

Includes the same additional features as Windows 98. However, it includes additional fixes and all of Year 2000 patches have been included in Windows 98 Second Edition. Below is a listing of the various new features Windows 98 SE includes:

Updates - Includes all the latest updates such as the Year 2000 updates, USB support, and more.

Software - Updated versions of Internet Explorer, Outlook Express, Windows Media Player, DirectX, NetMeeting and more.

Drivers / Support - Additional support for latest technology such as MMX, USB, Intel Pentium III, and more.

Computer Hope, Microsoft Windows 98. Retrieved January 20, 2012, from http://www.computerhope.com/win98.htm.

Microsoft Windows ME

Windows Millennium, also known as **Windows ME**, was introduced September 14, 2000, to the general public as the upgrade for Windows 95 and Windows 98 users, and is designed for end users. Overall, Windows ME has the look and feel of Windows 98 with some additional fixes and features not available in previous operating systems.

While Windows ME includes some of the latest fixes and updates and some enticing new features, we recommend this update only for users that may find or want some of the new features listed below or for users who are purchasing a new computer with this operating system.

Computer Hope, Microsoft Windows ME. Retrieved January 22, 2012, from http://www.computerhope.com/winme.htm.

Microsoft Windows NT

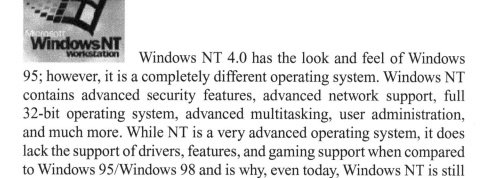

Windows NT 4.0 has the look and feel of Windows 95; however, it is a completely different operating system. Windows NT contains advanced security features, advanced network support, full 32-bit operating system, advanced multitasking, user administration, and much more. While NT is a very advanced operating system, it does lack the support of drivers, features, and gaming support when compared to Windows 95/Windows 98 and is why, even today, Windows NT is still used primarily by businesses and technical users.

Computer Hope, Microsoft Windows NT. Retrieved January 22, 2012, from http://www.computerhope.com/winnt.htm.

Microsoft Windows 2000 Professional

Windows 2000 is based on the Windows NT Kernel and is sometimes referred to as Windows NT 5.0. Windows 2000 contains over twenty-nine million lines of code, mainly written in C++.

Eight million of those lines alone are written for drivers. Currently, Windows 2000 is by far one of the largest commercial projects ever built.

Some of the significant features of Windows 2000 Professional are:

- Support for FAT16, FAT32, and NTFS.
- Increased uptime of the system and significantly fewer OS reboot scenarios.
- Windows Installer tracks applications and recognizes and replaces missing components.
- Protects memory of individual apps and processes to avoid a single app bringing the system down.
- Encrypting File Systems protects sensitive data.
- Secure Virtual Private Networking (VPN) supports tunneling in to private LAN over public Internet.
- Personalized menus adapt to the way you work.
- Multilingual version allows for User Interface and help to switch, based on logon.
- Includes broader support for high-speed networking devices, including Native ATM and cable modems.
- Supports USB and IEEE 1394 for greater bandwidth devices.

Microsoft Windows 2000 Server

Some of the significant features of Windows 2000 Server are:

- Active Directory improves manageability, enables security, and extends interoperability with other operating systems.
- Provides high-level interfaces for database access and Active Directory services.

- Lets you use COM+ to run component-based applications, integrated Web applications and message-queuing services.
- Transaction Services feature makes it easier to develop and deploy server-centric applications.
- Microsoft BackOffice is fully integrated into Windows 2000 Server.

Microsoft Windows 2000 Advanced Server

Some of the significant features of Windows 2000 Advanced Server are:

- The server operating system for e-commerce and line-of-business applications.
- Includes all the features of Windows 2000 Server, with additional scalability and clustering support.
- Increased reliability—ensure your business-critical applications are online when your customers need them
- Easier to use and manage clusters, applications, and updates
- Supports eight-way symmetric multiprocessing (SMP) and up to 8 GB of memory (RAM).

Computer Hope, Microsoft Windows 2000. Retrieved January 22, 2012, from http://www.computerhope.com/win2000.htm.

Microsoft Windows XP

Codenamed Whistler, Microsoft Windows XP is short for Windows Experienced and is the convergence of the two major Microsoft operating systems into one.

Windows XP is available in the below versions:

Home Edition - Full / Upgrade
Professional - Full / Upgrade

Windows XP is designed more for users who may not be familiar with all of Windows features and has several new abilities to make the Windows experience easier for those users.

Windows XP includes various new features not found in previous versions of Microsoft Windows. Below is a listing of some of these new features:

- New interface - a completely new look and ability to change the look.
- Updates - new feature that automatically obtains updates from the Internet.
- Internet Explorer 6 - includes Internet Explorer 6 and new IM.
- Multilingual support - added support for different languages.

In addition to the above features, Windows XP does increase reliability when compared to previous versions of Microsoft Windows.

Computer Hope, Microsoft Windows XP. Retrieved January 22, 2012, from http://www.computerhope.com/winxp.htm.

Microsoft Windows Vista

Microsoft Windows Vista is an upgrade to Microsoft Windows XP and Windows 2000 users. Windows Vista contains a dramatic new look for users used to previous versions of Microsoft Windows that has been designed to help create an overall better experience.

Some of the key new features of Window Vista include, among others:

- Windows Aero, a completely new GUI and unlike any previous version of Windows.
- Basic file backup and restore.
- Improved DVD support with the ability to easily create custom DVD movies.
- Easy transfer, a feature that allows you to easily transfer files from an older computer to the new computer.
- File encryption.
- Instant search available through all Explorer windows.
- Support for DirectX 10.
- Self-healing, the ability to automatically detect and correct problems that may be encountered on the computer.
- Shadow copy, a feature that allows you to recover deleted files.
- Improved photo gallery and control of photographs.
- Windows Sidebar and gadgets that allow you to add an almost endless list of different gadgets.
- More parental control.
- Improved Windows Calendar, with the ability to set tasks and appointments.

Computer Hope, Microsoft Windows Vista. Retrieved January 22, 2012, from http://www.computerhope.com/software/vista.htm.

Know Windows/Know your keyboard

Throughout the process of attaining your computer knowledge, or just being a seasoned computer user, there is one very important element you may want to consider—knowing your keyboard well. The theory and practice of knowing your keyboard cannot be separated from knowing the desktop functions of the operating system you're working within—in our case, Microsoft Windows.

Here below is a table of keyboard shortcut keys and the versions of Microsoft Windows they apply to, retrieved from computerhope.com. The table provides a summary of the individual versions of Windows and their applicable keyboard shortcuts. A second table below provides information on the functions of the Windows key, represented by "WINKEY." Please read.

Microsoft Windows shortcut keys

Below is a listing of all the major Windows shortcut keys and the versions of Microsoft Windows they are supported in.

Shortcut Keys	3.x	9x	ME	NT	2K	XP	Vista	7	Description
Alt + Tab	X	X	X	X	X	X	X	X	Switches between open applications
Alt + Shift + Tab	X	X	X	X	X	X	X	X	Switches backward between open applications
Alt + double-click		X	X	X	X	X	X	X	Displays the properties of the object you double-click on. For example, doing this on a file would display its properties

Command									Description
Ctrl + Tab	X	X	X	X	X	X	X	X	Switches between program groups or document windows in applications that support this feature
Ctrl + Shift + Tab	X	X	X	X	X	X	X	X	Same as above, but backward
Alt + Print Screen	X	X	X	X	X	X	X	X	Creates a screenshot only for the program you are currently in
Ctrl + Print Screen					X	X	X	X	Creates a screenshot of the entire screen
Ctrl + Alt + Del	X	X	X	X	X	X	X	X	Reboots the computer and brings up the Windows Task Manager
Ctrl + Shift + Esc					X	X	X	X	Immediately brings up the Windows Task Manager
Ctrl + Esc	X	X	X	X	X	X	X	X	Brings up the Windows Start menu. In Windows 3.x, this would bring up the Task Manager
Alt + Esc		X	X	X	X	X	X	X	Switches between open applications on the taskbar
F1	X	X	X	X	X	X	X	X	Activates Help for current open application
F2	X	X	X	X	X	X	X	X	Renames selected icon
F3		X	X	X	X	X	X	X	Starts Find from desktop
F4		X	X	X	X	X	X	X	Opens the drive selection when browsing
F5			X	X	X	X	X	X	Refreshes contents to where you were on the page
Ctrl + F5							X	X	Refreshes page to the beginning of the page
F10	X	X	X	X	X	X	X	X	Activates menu bar

Shortcut									Description
Shift + F10		X	X	X	X	X	X	X	Simulates right-click on selected item
F4		X	X	X	X	X	X	X	Select a different location to browse in the Windows Explorer toolbar
Alt + F4	X	X	X	X	X	X	X	X	Closes current open program
Ctrl + F4	X	X	X	X	X	X	X	X	Closes window in program
F6		X	X	X	X	X	X	X	Moves cursor to different Windows Explorer pane
Alt + Space bar	X	X	X	X	X	X	X	X	Drops down the window control menu
Ctrl + (the '+' key on the keypad)			X	X	X	X	X	X	Automatically adjusts the widths of all the columns in Windows Explorer
Alt + Enter		X	X	X	X	X	X	X	Opens Properties window of selected icon or program
Alt + Space bar		X	X	X	X	X	X	X	Opens the control menu for the current window open
Shift + Del		X	X	X	X	X	X	X	Deletes programs/files without throwing them into the Recycle Bin
Holding Shift	X	X	X	X	X	X	X	X	Boots in Safe Mode or bypasses system files as the computer is booting
Holding Shift		X	X	X	X	X	X	X	When putting in an audio CD, will prevent CD player from playing
Enter	X	X	X	X	X	X	X	X	Activates the highlighted program
Alt + Down arrow		X	X	X	X	X	X	X	Displays all available options on drop down menu

* (on the keypad)	X	X	X	X	X	X	X	Expands **all** folders on the currently selected folder or drive in Windows Explorer
+ (on the keypad)	X	X	X	X	X	X	X	Expands only the currently selected folder in Windows Explorer
- (on the keypad)	X	X	X	X	X	X	X	Collapses the currently selected folder in Windows Explorer

Windows keyboard shortcuts

Below is a listing of Windows keys that can be used on computers running a Microsoft Windows operating system and using a keyboard with a Windows key. In the below list of shortcuts, the Windows key is represented by "WINKEY."

Shortcut Keys	Description
WINKEY	Pressing the Windows key alone will open Start
WINKEY + F1	Opens the Microsoft Windows Help and Support Center
WINKEY + F3	Opens the Advanced Find window in Microsoft Outlook
WINKEY + D	Brings the desktop to the top of all other windows
WINKEY + M	Minimizes all windows
WINKEY + Shift + M	Undoes the minimize done by WINKEY + M and WINKEY + D
WINKEY + E	Opens Microsoft Explorer
WINKEY + Tab	Cycles through open programs through the taskbar
WINKEY + F	Displays the Windows Search/Find feature
WINKEY + Ctrl + F	Displays the Search for Computers window

WINKEY + R	Opens the Run window
WINKEY + Pause / Break key	Opens the System Properties Window
WINKEY + U	Opens Utility Manager
WINKEY + L	Locks the computer and switches users if needed (Windows XP and above only)
WINKEY + P	Quickly changes between monitor display types (Windows 7 only)
WINKEY + Left Arrow	Shrinks the window to half screen on the left side for side-by-side viewing (Windows 7 only)
WINKEY + Right Arrow	Shrinks the window to half screen on the right side for side-by-side viewing (Windows 7 only)
WINKEY + Up Arrow	When in the side-by-side viewing mode, this shortcut takes the screen back to full size (Windows 7 only)
WINKEY + Down Arrow	Minimizes the screen; Also, when in the side-by-side viewing mode, this shortcut takes the screen back to a minimized size (Windows 7 only)

Computer Hope, Microsoft Windows shortcut keys. Retrieved April 17, 2012, from http://www.computerhope.com/shortcut/windows.htm.

To break down the keyboard functions used in the Windows operating system seen above, I turned to Microsoft.com for more information. I retrieved a few tables and content that individualize the keys and key combinations and what they perform on the keyboard. Please read on.

General keyboard shortcuts

The following table contains general keyboard shortcuts:

Press this key	To do this
F1	Display Help
Ctrl+C (or Ctrl+Insert)	Copy the selected item
Ctrl+X	Cut the selected item
Ctrl+V (or Shift+Insert)	Paste the selected item
Ctrl+Z	Undo an action
Ctrl+Y	Redo an action
Delete (or Ctrl+D)	Delete the selected item and move it to the Recycle Bin
Shift+Delete	Delete the selected item without moving it to the Recycle Bin first
F2	Rename the selected item
Ctrl+Right Arrow	Move the cursor to the beginning of the next word
Ctrl+Left Arrow	Move the cursor to the beginning of the previous word
Ctrl+Down Arrow	Move the cursor to the beginning of the next paragraph
Ctrl+Up Arrow	Move the cursor to the beginning of the previous paragraph
Ctrl+Shift with an arrow key	Select a block of text
Shift with any arrow key	Select more than one item in a window or on the desktop, or select text within a document
Ctrl with any arrow key+Space bar	Select multiple individual items in a window or on the desktop
Ctrl+A	Select all items in a document or window
F3	Search for a file or folder
Alt+Enter	Display properties for the selected item

Press this key	To do this
Alt+F4	Close the active item, or exit the active program
Alt+Space bar	Open the shortcut menu for the active window
Ctrl+F4	Close the active document (in programs that allow you to have multiple documents open simultaneously)
Alt+Tab	Switch between open items
Ctrl+Alt+Tab	Use the arrow keys to switch between open items
Ctrl+Mouse scroll wheel	Change the size of icons on the desktop
Windows logo key ⊞ +Tab	Cycle through programs on the taskbar by using Aero Flip 3-D
Ctrl+Windows logo key ⊞ +Tab	Use the arrow keys to cycle through programs on the taskbar by using Aero Flip 3-D
Alt+Esc	Cycle through items in the order in which they were opened
F6	Cycle through screen elements in a window or on the desktop
F4	Display the address bar list in Windows Explorer
Shift+F10	Display the shortcut menu for the selected item
Ctrl+Esc	Open the Start menu
Alt+underlined letter	Display the corresponding menu
Alt+underlined letter	Perform the menu command (or other underlined command)
F10	Activate the menu bar in the active program
Right Arrow	Open the next menu to the right, or open a submenu
Left Arrow	Open the next menu to the left, or close a submenu
F5 (or Ctrl+R)	Refresh the active window

Press this key	To do this
Alt+Up Arrow	View the folder one level up in Windows Explorer
Esc	Cancel the current task
Ctrl+Shift+Esc	Open Task Manager
Shift when you insert a CD	Prevent the CD from automatically playing
Left Alt+Shift	Switch the input language when multiple input languages are enabled
Ctrl+Shift	Switch the keyboard layout when multiple keyboard layouts are enabled
Right or Left Ctrl+ Shift	Change the reading direction of text in right-to-left reading languages

Microsoft Corporation, General keyboard shortcuts. Retrieved March 17, 2012, from http://windows.microsoft.com/en-US/Windows7/Keyboard-shortcuts.

Windows Logo keyboard shortcuts

The following table contains keyboard shortcuts that use the Windows logo key ⊞.

Press this key	To do this
Windows logo key ⊞	Open or close the Start menu
Windows logo key ⊞ +Pause	Display the System Properties dialog box
Windows logo key ⊞+D	Display the desktop
Windows logo key ⊞+M	Minimize all windows
Windows logo key ⊞ +Shift+M	Restore minimized windows to the desktop
Windows logo key ⊞+E	Open Computer
Windows logo key ⊞+F	Search for a file or folder

Press this key	To do this
Ctrl+Windows logo key ⊞+F	Search for computers (if you're on a network)
Windows logo key ⊞+L	Lock your computer or switch users
Windows logo key ⊞+R	Open the Run dialog box
Windows logo key ⊞+T	Cycle through programs on the taskbar
Windows logo key ⊞ +number	Start the program pinned to the taskbar in the position indicated by the number. If the program is already running, switch to that program
Shift+Windows logo key ⊞+number	Start a new instance of the program pinned to the taskbar in the position indicated by the number
Ctrl+Windows logo key ⊞+number	Switch to the last active window of the program pinned to the taskbar in the position indicated by the number
Alt+Windows logo key ⊞+number	Open the Jump List for the program pinned to the taskbar in the position indicated by the number
Windows logo key ⊞ +Tab	Cycle through programs on the taskbar by using Aero Flip 3-D
Ctrl+Windows logo key ⊞+Tab	Use the arrow keys to cycle through programs on the taskbar by using Aero Flip 3-D
Ctrl+Windows logo key ⊞+B	Switch to the program that displayed a message in the notification area
Windows logo key ⊞ +Space bar	Preview the desktop
Windows logo key ⊞+ Up Arrow	Maximize the window
Windows logo key ⊞ +Left Arrow	Maximize the window to the left side of the screen
Windows logo key ⊞ +Right Arrow	Maximize the window to the right side of the screen
Windows logo key ⊞ +Down Arrow	Minimize the window

Press this key	To do this
Windows logo key ⊞ +Home	Minimize all but the active window
Windows logo key ⊞ +Shift+Up Arrow	Stretch the window to the top and bottom of the screen
Windows logo key ⊞ +Shift+Left Arrow or Right Arrow	Move a window from one monitor to another
Windows logo key ⊞+P	Choose a presentation display mode
Windows logo key ⊞+G	Cycle through gadgets
Windows logo key ⊞+U	Open Ease of Access Center
Windows logo key ⊞+X	Open Windows Mobility Center

Microsoft Corporation, Windows Logo keyboard shortcuts. Retrieved March, 17, 2012, from http://windows.microsoft.com/en-US/Windows7/Keyboard-shortcuts.

Windows Explorer keyboard shortcuts

The following table contains keyboard shortcuts for working with Windows Explorer windows or folders.

Press this key	To do this
Ctrl+N	Open a new window
Ctrl+W	Close the current window
Ctrl+Shift+N	Create a new folder
End	Display the bottom of the active window
Home	Display the top of the active window
F11	Maximize or minimize the active window
Ctrl+Period (.)	Rotate a picture clockwise
Ctrl+Comma (,)	Rotate a picture counter-clockwise
Num Lock+Asterisk (*) on numeric keypad	Display all subfolders under the selected folder

Press this key	To do this
Num Lock+Plus Sign (+) on numeric keypad	Display the contents of the selected folder
Num Lock+Minus Sign (-) on numeric keypad	Collapse the selected folder
Left Arrow	Collapse the current selection (if it's expanded), or select the parent folder
Alt+Enter	Open the Properties dialog box for the selected item
Alt+P	Display the preview pane
Alt+Left Arrow	View the previous folder
Backspace	View the previous folder
Right Arrow	Display the current selection (if it's collapsed), or select the first subfolder
Alt+Right Arrow	View the next folder
Alt+Up Arrow	View the parent folder
Ctrl+Shift+E	Display all folders above the selected folder
Ctrl+Mouse scroll wheel	Change the size and appearance of file and folder icons
Alt+D	Select the address bar
Ctrl+E	Select the search box
Ctrl+F	Select the search box

Microsoft Corporation, Windows Explorer keyboard shortcuts. Retrieved March, 17, 2012, from http://windows.microsoft.com/en-US/ Windows7/Keyboard-shortcuts.

Windows Vista

Microsoft keyboard shortcuts

The following table contains keyboard shortcuts for use with Microsoft keyboards.

Press this key	To do this
Windows logo key ⊞	Open or close the Start menu
Windows logo key ⊞ +PAUSE	Display the System Properties dialog box
Windows logo key ⊞ +D	Display the desktop
Windows logo key ⊞ +M	Minimize all windows
Windows logo key ⊞ +Shift+M	Restore minimized windows to the desktop
Windows logo key ⊞ +E	Open Computer
Windows logo key ⊞ +F	Search for a file or folder
Ctrl+Windows logo key ⊞+F	Search for computers (if you are on a network)
Windows logo key ⊞ +L	Lock your computer or switch users
Windows logo key ⊞ +R	Open the Run dialog box
Windows logo key ⊞ +T	Cycle through programs on the taskbar
Windows logo key ⊞ +Tab	Cycle through programs on the taskbar by using Windows Flip 3-D
Ctrl+Windows logo key ⊞+Tab	Use the arrow keys to cycle through programs on the taskbar by using Windows Flip 3-D
Windows logo key ⊞ +Space bar	Bring all gadgets to the front and select Windows Sidebar
Windows logo key ⊞ +G	Cycle through Sidebar gadgets
Windows logo key ⊞ +U	Open Ease of Access Center
Windows logo key ⊞ +X	Open Windows Mobility Center

Press this key	To do this
Windows logo key ⊞ with any number key	Open the Quick Launch shortcut that is in the position that corresponds to the number. For example, Windows logo key ⊞+1 to launch the first shortcut in the Quick Launch menu.

Microsoft Corporation, Microsoft keyboard shortcuts. Retrieved March 17, 2012, from http://windows.microsoft.com/en-US/windows-vista/Keyboard-shortcuts.

Windows XP

Keyboard Shortcuts for Windows XP

To	Press
Set focus on a notification	**Windows Key+B**
View properties for the selected item	**Alt+Enter**
Displays the properties of the selected object	**Alt+Enter**
Cycle through items in the order they were opened	**Alt+Esc**
Close the active item, or quit the active program	**Alt+F4**
Opens the shortcut menu for the active window	**Alt+Space bar**
Display the System menu for the active window	**Alt+Space bar**
Switch between open items	**Alt+Tab**
Carry out the corresponding command or select the corresponding option in a dialog box	**Alt+underlined letter**
Display the corresponding menu	**Alt+underlined letter in a menu name**
Select a button if the active option is a group of option buttons in a dialog box	**Arrow keys**

To	Press
View the folder one level up in My Computer or Windows Explorer	**Backspace**
Open a folder one level up if a folder is selected in the Save As or Open dialog box in a dialog box	**Backspace**
Copy selected item	**Ctrl while dragging an item**
Select all	**Ctrl+A**
Copy	**Ctrl+C**
Move the insertion point to the beginning of the next paragraph	**Ctrl+Down Arrow**
Display the Start menu	**Ctrl+Esc**
Close the active document in programs that allow you to have multiple documents open simultaneously	**Ctrl+F4**
Move the insertion point to the beginning of the previous word	**Ctrl+Left Arrow**
Move the insertion point to the beginning of the next word	**Ctrl+Right Arrow**
Create shortcut to selected item	**Ctrl+Shift while dragging an item**
Highlight a block of text	**Ctrl+Shift with any of the arrow keys**
Move backward through tabs in a dialog box	**Ctrl+Shift+Tab**
Move forward through tabs in a dialog box	**Ctrl+Tab**
Move the insertion point to the beginning of the previous paragraph	**Ctrl+Up Arrow**
Paste	**Ctrl+V**
Search for computers	**Ctrl+Windows Key+F**
Cut	**Ctrl+X**
Undo	**Ctrl+Z**

To	Press
Delete	**Delete**
Display the bottom of the active window	**End**
Carry out the command for the active option or button in a dialog box	**Enter**
Cancel the current task	**Esc**
Display Help in a dialog box	**F1**
Activate the menu bar in the active program	**F10**
Rename selected item	**F2**
Search for a file or folder	**F3**
Display the Address bar list in My Computer or Windows Explorer	**F4**
Display the items in the active list in a dialog box	**F4**
Refresh the active window	**F5**
Cycle through screen elements in a window or on the desktop	**F6**
Display the top of the active window	**Home**
Switch MouseKeys on and off	**Left Alt+left Shift+Num Lock**
Switch High Contrast on and off	**Left Alt+left Shift+Print Screen**
Open the next menu to the left, or close a submenu	**Left Arrow**
Collapse current selection if it's expanded, or select parent folder	**Left Arrow**
Display the shortcut menu for the selected item	**Menu key**
Switch ToggleKeys on and off	**Num Lock for five seconds**
Display all subfolders under the selected folder	**Num Lock+asterisk on a numeric keypad (*)**

To	Press
Collapse the selected folder	**Num Lock+minus sign on a numeric keypad (-)**
Display the contents of the selected folder	**Num Lock+plus sign on a numeric keypad (+)**
Open the next menu to the right, or open a submenu	**Right Arrow**
Display current selection if it's collapsed, or select first subfolder	**Right Arrow**
Switch FilterKeys on and off	**Right Shift for eight seconds**
Switch StickyKeys on and off	**Shift five times**
Prevent the CD from automatically playing	**Shift when you insert a CD into the CD-ROM drive**
Select more than one item in a window or on the desktop, or select text within a document	**Shift with any of the arrow keys**
Delete selected item permanently without placing the item in the Recycle Bin	**Shift+Delete**
Display the shortcut menu for the selected item	**Shift+F10**
Move backward through options in a dialog box	**Shift+Tab**
Select or clear the check box if the active option is a check box in a dialog box	**Space bar**
Move forward through options in a dialog box	**Tab**
Carry out the corresponding command	**Underlined letter in a command name on an open menu**
Display or hide the Start menu	**Windows Key**

To	Press
Lock your computer if you are connected to a network domain, or switch users if you are not connected to a network domain	**Windows Key+L**
Display the System Properties dialog box	**Windows Key+Break**
Show the desktop	**Windows Key+D**
Open My Computer	**Windows Key+E**
Search for a file or folder	**Windows Key+F**
Display Windows Help	**Windows Key+F1**
Minimize all windows	**Windows Key+M**
Open the Run dialog box	**Windows Key+R**
Restores minimized windows	**Windows Key+Shift+M**
Opens Utility Manager	**Windows Key+U**

Microsoft Corporation, keyboard Shortcuts for Windows XP. Retrieved March 17, 2012, from http://www.microsoft.com/enable/products/KeyboardSearch_XP.aspx.

Internet Explorer 9 keyboard shortcuts

You can use Windows Internet Explorer shortcut keys to perform lots of different tasks quickly, or to work without a mouse.

Viewing and exploring Web pages

The following table describes shortcuts used to view and explore Web pages.

To do this	Press this
Display Help	F1
Toggle between full-screen and regular views of the browser window	F11

Move forward through the items on a Web page, the Address bar, or the Favorites bar	Tab
Move back through the items on a Web page, the Address bar, or the Favorites bar	Shift+Tab
Start Caret Browsing	F7
Go to your home page	Alt+Home
Go to the next page	Alt+Right Arrow
Go to the previous page	Alt+Left Arrow or Backspace
Display a shortcut menu for a link	Shift+F10
Move forward through frames and browser elements (only works if tabbed browsing is turned off)	Ctrl+Tab or F6
Move backward between frames (only works if tabbed browsing is turned off)	Ctrl+Shift+Tab
Scroll toward the beginning of a document	Up Arrow
Scroll toward the end of a document	Down Arrow
Scroll toward the beginning of a document in larger increments	Page Up
Scroll toward the end of a document in larger increments	Page Down
Move to the beginning of a document	Home
Move to the end of a document	End
Find on this page	Ctrl+F
Refresh the current Web page	F5
Refresh the current Web page, even if the time stamp for the Web version and your locally stored version are the same	Ctrl+F5
Stop downloading a page	Esc
Open a new Web site or page	Ctrl+O
Open a new window	Ctrl+N
Open a new InPrivate Browsing window	Ctrl+Shift+P
Delete browsing history	Ctrl+Shift+Delete

Duplicate tab (open current tab in a new tab)	Ctrl+K
Reopen the last tab you closed	Ctrl+Shift+T
Close the current window (if you only have one tab open)	Ctrl+W
Save the current page	Ctrl+S
Print the current page or active frame	Ctrl+P
Activate a selected link	Enter
Open Favorites	Ctrl+I
Open History	Ctrl+H
Open Download Manager	Ctrl+J
Open the Page menu (if the Command bar is visible)	Alt+P
Open the Tools menu (if the Command bar is visible)	Alt+T
Open the Help menu (if the Command bar is visible)	Alt+H

Microsoft Corporation, Internet Explorer 9 keyboard shortcuts. Retrieved April 17, 2012, from http://windows.microsoft.com/en-US/windows7/Internet-Explorer-9-keyboard-shortcuts.

Windows Photo Gallery keyboard shortcut

The following table contains keyboard shortcuts for working with Windows Photo Gallery:

Press this key	To do this
Ctrl+F	Open the Fix pane
Ctrl+P	Print the selected picture
Enter	View the selected picture at a larger size
Ctrl+I	Open or close the Details pane
Ctrl+period (.)	Rotate the picture clockwise
Ctrl+comma (,)	Rotate the picture counter-clockwise

Press this key	To do this
F2	Rename the selected item
Ctrl+E	Search for an item
Alt+Left Arrow	Go back
Alt+Right Arrow	Go forward
Plus sign (+)	Zoom in or resize the picture thumbnail
Minus sign (-)	Zoom out or resize the picture thumbnail
Ctrl+Mouse scroll wheel	Change the size of the picture thumbnail
Ctrl+B	Best fit
Left Arrow	Select the previous item
Down Arrow	Select the next item or row
Up Arrow	Previous item (Easel) or previous row (Thumbnail)
Page Up	Previous screen
Page Down	Next screen
Home	Select the first item
End	Select the last item
Delete	Move the selected item to the Recycle Bin
Shift+Delete	Permanently delete the selected item
Left Arrow	Collapse node
Right Arrow	Expand node

Keyboard shortcuts for working with videos

J	Move back one frame
K	Pause the playback
L	Move forward one frame
I	Set the start trim point
O	Set the end trim point
M	Split a clip

Press this key	To do this
Home	Stop and rewind all the way back to the start trim point
Alt+Right Arrow	Advance to the next frame
Alt+Left Arrow	Go back to the previous frame
Ctrl+K	Stop and rewind playback
Ctrl+P	Play from the current location
Home	Move the start trim point
End	Move to the end trim point
Page Up	Seek to nearest split point before the current location
Page Down	Seek to nearest split point after the current location

Microsoft Corporation, Windows Photo Gallery keyboard shortcuts. Retrieved April 17, 2012, from http://windows.microsoft.com/en-US/ windows-vista/Keyboard-shortcuts.

Using your keyboard

Whether you're writing a letter or entering numerical data, your keyboard is the main way to enter information into your computer. But did you know you can also use your keyboard to control your computer? Learning just a few simple keyboard commands (instructions to your computer) can help you work more efficiently. This article covers the basics of keyboard operation and gets you started with keyboard commands.

How the keys are organized

The keys on your keyboard can be divided into several groups based on function:

- **Typing (alphanumeric) keys.** These keys include the same letter, number, punctuation, and symbol keys found on a traditional typewriter.

- **Control keys.** These keys are used alone or in combination with other keys to perform certain actions. The most frequently used control keys are Ctrl, Alt, the Windows key *※*, and Esc.
- **Function keys.** The function keys are used to perform specific tasks. They are labeled as F1, F2, F3, and so on, up to F12. The functionality of these keys differs from program to program.
- **Navigation keys.** These keys are used for moving around in documents or Web pages and editing text. They include the arrow keys, Home, End, Page Up, Page Down, Delete, and Insert.
- **Numeric keypad.** The numeric keypad is handy for entering numbers quickly. The keys are grouped together in a block, like a conventional calculator or adding machine.

The following illustration shows how these keys are arranged on a typical keyboard. Your keyboard layout may differ.

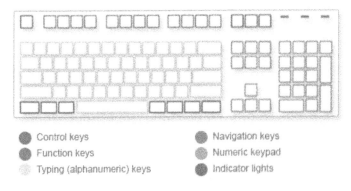

Control keys   Navigation keys
Function keys   Numeric keypad
Typing (alphanumeric) keys   Indicator lights

How the keys are arranged on a keyboard

Typing text

Whenever you need to type something in a program, e-mail message, or text box, you'll see a blinking vertical line (|). That's the cursor, also called the insertion point. It shows where the text that you type will begin. You can move the cursor by clicking on the desired location with the mouse, or by using the navigation keys (see the "Using navigation keys" section of this article).

In addition to letters, numerals, punctuation marks, and symbols, the typing keys also include Shift, Caps Lock, the Tab key, Enter, the Space bar, and Backspace.

Key name	How to use it
Shift	Press Shift in combination with a letter to type an uppercase letter. Press Shift in combination with another key to type the symbol shown on the upper part of that key.
Caps Lock	Press Caps Lock once to type all letters as uppercase. Press Caps Lock again to turn this function off. Your keyboard may have a light indicating whether Caps Lock is on.
Tab	Press the Tab key to move the cursor several spaces forward. You can also press the Tab key to move to the next text box on a form.
Enter	Press Enter to move the cursor to the beginning of the next line. In a dialog box, press Enter to select the highlighted button.
Space bar	Press the Space bar to move the cursor one space forward.
Backspace	Press Backspace to delete the character before the cursor, or the selected text.

Using keyboard shortcuts

Keyboard shortcuts are ways to perform actions by using your keyboard. They're called shortcuts because they help you work faster. In fact, almost any action or command you can perform with a mouse can be performed faster using one or more keys on your keyboard.

In Help topics, a plus sign (+) between two or more keys indicates that those keys should be pressed in combination. For example, Ctrl+A means to press and hold Ctrl and then press A. Ctrl+Shift+A means to press and hold Ctrl and Shift and then press A.

Find program shortcuts

You can do things in most programs by using the keyboard. To see which commands have keyboard shortcuts, open a menu. The shortcuts (if available) are shown next to the menu items.

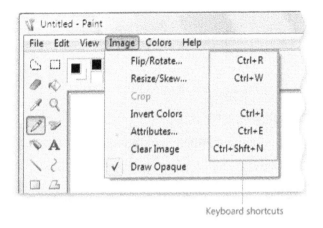

Keyboard shortcuts

Choose menus, commands, and options

You can open menus and choose commands and other options using your keyboard. When you press Alt in a program with menus, one letter in each of the menu names becomes underlined. Press an underlined letter to open the corresponding menu. Press the underlined letter in a menu item to choose that command.

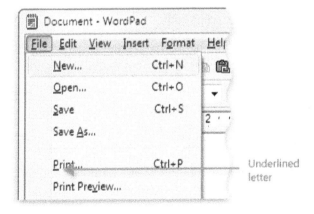

Underlined letter

Press Alt+F to open the File menu, then press P to choose the Print command

This trick works in dialog boxes too. Whenever you see an underlined letter attached to an option in a dialog box, it means you can press Alt plus that letter to choose that option.

Useful shortcuts

The following table lists some of the most useful keyboard shortcuts.

Press this	To do this
Windows logo key ⊞	Open the Start menu
Alt+Tab	Switch between open programs or windows
Alt+F4	Close the active item, or exit the active program
Ctrl+S	Save the current file or document (works in most programs)
Ctrl+C	Copy the selected item
Ctrl+X	Cut the selected item
Ctrl+V	Paste the selected item
Ctrl+Z	Undo an action
Ctrl+A	Select all items in a document or window
F1	Display Help for a program or Windows
Windows logo key ⊞+F1	Display Windows Help and Support
Esc	Cancel the current task
Application key ▤	Open a menu of commands related to a selection in a program. Equivalent to right-clicking the selection.

Using navigation keys

The navigation keys allow you to move the cursor, move around in documents and Web pages, and edit text. The following table lists some common functions of these keys.

Press this	To do this
Left Arrow, Right Arrow, Up Arrow, or Down Arrow	Move the cursor or selection one space or line in the direction of the arrow, or scroll a Web page in the direction of the arrow

Home	Move the cursor to the beginning of a line or move to the top of a Web page
End	Move the cursor to the end of a line or move to the bottom of a Web page
Ctrl+Home	Move to the top of a document
Ctrl+End	Move to the bottom of a document
Page Up	Move the cursor or page up one screen
Page Down	Move the cursor or page down one screen
Delete	Delete the character after the cursor, or the selected text; in Windows, delete the selected item and move it to the Recycle Bin
Insert	Turn Insert mode off or on. When Insert mode is on, text that you type is inserted at the cursor. When Insert mode is off, text that you type replaces existing characters.

Using the numeric keypad

The numeric keypad arranges the numerals 0 though 9, the arithmetic operators + (addition), - (subtraction), * (multiplication), and / (division), and the decimal point as they would appear on a calculator or adding machine. These characters are duplicated elsewhere on the keyboard, of course, but the keypad arrangement allows you to rapidly enter numerical data or mathematical operations with one hand.

Numeric keypad

To use the numeric keypad to enter numbers, press Num Lock. Most keyboards have a light that indicates whether Num Lock is on or off. When Num Lock is off, the numeric keypad functions as a second set of navigation keys (these functions are printed on the keys next to the numerals or symbols).

You can use your numeric keypad to perform simple calculations with Calculator.

Operate Calculator with the numeric keypad

1. Open Calculator by clicking the Start button, clicking *All Programs*, clicking *Accessories*, and then clicking *Calculator*.
2. Check your keyboard light to see if Num Lock is on. If it isn't, press Num Lock.
3. Using the numeric keypad, type the first number in the calculation.
4. On the keypad, type + to add, - to subtract, * to multiply, or / to divide.
5. Type the next number in the calculation.
6. Press Enter to complete the calculation.

Three odd keys

So far, we've discussed almost every key you're likely to use. But for the truly inquisitive, let's explore the three most mysterious keys on the keyboard: Print Screen, Scroll Lock, and Pause/Break.

PRINT SCREEN (or PRT SCN)

A long time ago, this key actually did what it says—it sent the current screen of text to your printer. Nowadays, pressing Print Screen captures an image of your entire screen (a "screenshot") and copies it to the Clipboard in your computer's memory. From there you can paste it (Ctrl+V) into Microsoft Paint or another program and, if you want, print it from that program.

More obscure is Sys Rq, which shares the key with Print Screen on some keyboards. Historically, Sys Rq was designed to be a "system request," but this command is not enabled in Windows.

Tip

Press Alt+Print Screen to capture an image of just the active window, instead of the entire screen.

SCROLL LOCK (or SCR LK)

In most programs, pressing Scroll Lock has no effect. In a few programs, pressing Scroll Lock changes the behavior of the arrow keys and the Page Up and Page Down keys; pressing these keys causes the document to scroll without changing the position of the cursor or selection. Your keyboard might have a light indicating whether Scroll Lock is on.

PAUSE/BREAK

This key is rarely used. In some older programs, pressing this key pauses the program or, in combination with Ctrl, stops it from running.

Microsoft Corporation, Using your keyboard. Retrieved February 16, 2012, from http://windows.microsoft.com/en-US/windows-vista/Using-your-keyboard.

What Is the Internet?

The Internet is a global system of interconnected computer networks. The Internet may also be defined as a global communications network, and it is considered as a network of all networks through which millions of networks (private, public, government, business) of all types and scopes are interconnected. All of these networks use the same type of communication protocol called Transmission Control Protocol/Internet Protocol (TCP/IP). The TCP/IP is a standard by which all Internet users must comply for the transmission of digital data.

A wide array of information resources transmit via the Internet, including billions of interconnected Web pages that are transferred using HTTP (Hypertext Transfer Protocol), and are collectively known as the World Wide Web. The Internet also uses File Transfer Protocol (FTP) to transfer files, and Simple Mail Transfer Protocol (SMTP) to transfer e-mail. Details of how these protocols execute Internet tasks are found in many sources of learning, including books, journals and the Internet itself.

Browsing the Internet

The more you become involved with the computer and computing in general, you will constantly hear terms related to many tasks that are performed locally on the computer and by other means. One of those tasks is using the Internet from a computer or an electronic device to establish the connection to access and communicate with another computer or device. To be allowed to do so, you first connect through a network which, in turn, has established a connection to the Internet. When navigating the Internet through an established connection you may consider yourself as browsing or surfing the Internet, the Web, or the Net, which are other simple terms used when referring to the Internet.

What Is an Internet Browser?

Think of the Web browser as the principal messenger for the Internet. According to Wikipedia, "a **Web browser** is a software application for retrieving, presenting, and traversing information resources on the

World Wide Web. An *information resource* is identified by a Uniform Resource Identifier (URI) and may be a Web page, image, video, or other piece of content. Hyperlinks present in resources enable users easily to navigate their browsers to related resources. A Web browser can also be defined as an application software or program designed to enable users to access, retrieve and view documents and other resources on the Internet."

Wikipedia, Web browser. Retrieved May 8, 2012, from http:// en.wikipedia.org/wiki/Web_browser.

The Browser and a Client/Server Scenario

With the client/server scenario in mind, one should think of the browser as the principle runner between the client and server, the server and server or the client and client. When running on the client machine, the browser contacts the server for resources as requested by the user/client. At that point, the server is the principal host of the needed resources. The server then uses server method to authenticate the request, and when approved, the server permits the browser to retrieve the requested information or resource. When retrieved, the browser returns to the networked computer bearing the information it has acquired from the server. This action is conducted from the side of the end user of the computer in the following manner: the user launches (opens) the browser from the computer; the user feeds into the Universal Resources Locator (URL) the server's address from which the information is needed and sets browser off with the request. The server at the other end receives the request and responds by providing the needed information upon authentication of the browser's source address. The browser returns to the source computer with the needed information.

History of the Web Browser

The first ever browser to the Internet was called the WorldWideWeb (no spaces), invented in 1990 by Tim Berners-Lee. The WorldWideWeb browser was later renamed Nexus. This invention was followed later in 1993 by the invention by Marc Andreesen of the Mosaic browser, later renamed Netscape Navigator by his company, Netscape. Ever since then, many browsers have emerged, and the Internet has become more revolutionized by the functions, power, and speed produced by newer browser types. Today (around the time of this book), the most popular Web browsers to the Internet are Microsoft Internet Explorer, Mozilla Firefox, Safari by Apple, Google Chrome, and Opera.

Using the Web Browser

The longer you browse the Internet and the more Web sites you visit, the more there are browser add-ons and Web offers by different software vendors/owners you come in contact with, not to mention various ads being pushed by various programs across the Net. Software vendors work with one another to promote their products in places frequented by Web visitors. Many times, you find yourself chasing after a free Web product (freeware), or simply something you think you can acquire at no cost to help fulfill your computing needs. During your search, you may be directed to a Web site or a host of linked Web sites, and you may be lucky to find what you're looking for, or not at all.

In most cases, you may find yourself entrapped by products you were not looking for. Some of them may be simply browser add-ons. Others are software intentionally pushed down through the Internet browser during your visit. Many of them will simply warn you to agree or not agree to their terms by simply clicking a radio button, even though they become sneaky about it in most cases. There are times, in fact, when you may be coaxed into establishing a profile or creating a personal account for easy access during your next Web visit. Sometimes, you feel you need the product or resource so badly that you must comply with the terms and establish a profile.

Your goal is to get past the hurdles and beat every obstacle in the way to ultimately get to the download of your sought-after free software. The next thing you know, you have completed your download operation, but as you return to your desktop you notice it saturated with some strange icons. You wonder what the icons are all about. Of course, they are icons of products that you do not need—they came with what you had just downloaded to your computer. Many of them configure themselves to run automatically and launch when you start your computer. They cause dramatic slowdowns to your computer system, but you can remove or uninstall them through the Control Panel.

Note: the Control Panel is an area from which you must be careful to perform computer tasks. You must be acquainted with each section of the panel to perform a task, and whatever program you attempt to uninstall. Some programs in the Control Panel are system programs preinstalled through the operating system, and they help the system perform properly. You do not want to uninstall such programs. In most cases, however, you may be lucky to get denied access, even at an administrative login, to uninstall some preinstalled system programs. The operating system is that smart to take such precautions to preserve these programs.

Also, the method of accessing the Control Panel is beyond the scope of this book, and the subject will not be discussed.

Browser Functions

There are numerous functions the Web browser can perform to facilitate a mission desired by the end user. This book will use three of the most popular browsers in use today to explain the standard functions of the Internet browser.

Back   Back   Back

The Back button: This button takes you back to the previous page you were at on the Web browser.

Forward Forward Forward

The Forward button: This button takes you to the next screen in your history list or to a page you had most previously launched. The button is grayed out unless you had launched another page once.

Refresh Reload Reload

The Refresh/Reload button: This button, when clicked, refreshes/reloads the browser page properly. Due to network problems or some sort of multitasking issue, the browser launch might have failed the first time. The function also works to reset browsers' pages that are timed out.

Home   Home   Home

The Home button: This takes you to the browser page that has been set as the home page.

Print   Print   Print

The Print button: This button sends a printing command to a printer attached to the computer that the user is working from.

Favorites/Bookmarks: A function/term used to save Web pages by the browser into a folder on the computer to be accessed at a later time.

Browser Tricks

1. **Visiting multiple sites:** You are on the Internet and have multiple Web sites to visit at one time. Would you open so many sessions of the Web browser and clutter up the taskbar? What would you do?

    Answer: Use the one session of a browser and open several tabs. The shortcut method to open a new tab on the browser is by holding Ctrl+T. Alternatively, you may click the blank area or the plus sign (+) at the top of the browser to open a new tab. Type in the Web address desired. The two methods above are in addition to opening a new tab traditionally from the "File" menu.

2. **Clean up your browser (clear the cache)**: You're on the Internet a lot. Have you made it a priority to clean up your browser from time to time? You should get in the habit of clearing the cache to help speed up your browser. From those many Web sites you last visited, your browser has stored too many temporary Internet files. Also, the cache contains details of your browsing history. There might be too many large files, including video, audio, and image files. How would you clear the cache now to help improve Internet speed?

    Answer: In Internet Explorer - Click the *Tools* menu and select *Internet Options* at the bottom. In the *Browsing History* area in the middle, click the *Delete* button. In the dialog box that opens, check the box at the type of files you wish to delete.

    In Firefox - Click *Tools* and select *Options* at the bottom. Click on *Advanced* to the far right and click the *Network* tab. Click the *Clear Now* button to clear the cache.

3. **Turn off applications**: Whether or not you have access to the Control Panel to uninstall unwanted applications, you want to utilize the speed of your computer and the Internet browser it runs. This slowdown issue may be caused by the computer itself as a result of unwanted apps loaded by the browser and installed onto your computer. As you look to the lower right hand corner at

the notification area, you see icons of apps you don't recognize. What do you do?

Answer: Right-click each icon you do not recognize and choose to exit from it, close it, or turn it off. It may still be a useful application, but you do not need it running at this time. It is taking up memory space and slowing down your computer. Make sure you do not turn off useful applications.

4. **Disable Add-ons**: Add-ons are retrieved from the Internet. Add-ons are intended by some sites to make your Web experience more interesting. In most cases, add-ons provide enhanced features for visual or audio reception by the Internet user. But in some cases, add-ons are pushed onto your computer browser without your consent or knowledge, and by the time you are finished with a session of Internet browsing, your browser is full. The programs automatically get installed, and you can see all the space taken up on the browser toolbar. Of course, you may notice that your computer's slowdown has begun to take gradual effect.

    Whether preinstalled by the computer manufacturer or downloaded at one point during the computer's use, you cannot delete add-ons of any sort; you can only disable them.

    This is how you can disable add-ons in Internet Explorer 8 and later using Add-on Manager:

    • Launch (open) Internet Explorer from the Start menu.
    • Click the Tools menu and choose *Internet Options*.
    • Click the *Programs* tab to the top and click the *Manage Add-ons* button in the middle.
    • In the *Manage Add-ons* dialog box, select (click) the one you wish to disable.

    If, before you disable any add-ons, you would want a clean browser with no add-ons enabled and try launching it from the System Tools directory, this is how:

Click the Start button then click *All Programs*. Under that, click *Accessories*, then click *System Tools*. Under there, click *Internet Explorer (No Add-ons)*.

If you are working from Mozilla Firefox, below is how you can disable add-ons:

- Click the Tools menu to the top and click *Options*.
- In the Options dialog box, click the *Manage Add-ons* button.
- Here, you can enable/disable plugins, extensions, and add-ons as needed.

Bonus Tip

When you find time to launch your browser and surf the Internet, you don't want anything to stop you, not even the browser itself. But there may be times when your frustrations almost reach the peak, and you feel like the computer is just not working. More often than not, the computer seems to have frozen, stalled, or simply stopped working when the browser is saturated with add-ons from all over the Web.

Aside from tips provided earlier, such as clearing the cache and performing functions that will help the browser regain full functionality, my advice for the seasoned Web enthusiast is that you always should have a second browser ready to use in the event your default one stops working properly. I personally am a huge Microsoft Internet Explorer (IE) fan, and I recognize that IE can be exceptionally prone to freezing. This happens mostly when much-needed updates to computer resources do not take place when due, or when some updates are missed and do not take place at all. As such, there are chances that there would be a conflict between computer registry settings and computer resource settings, and that causes miscommunication in the computer's operating system. That, in turn, creates a big setback with the operating system, resulting into a significant slowdown in the computer's functionality, much of it happening through browsing activities.

If you use Internet Explorer quite often in the Windows computing environment, here below are links you can follow to download and install at least one additional browser for backup:

> For Mozilla Firefox - Visit www.mozilla.org, then select your region and follow the instructions to download and install.
> For Google Chrome - Visit www.google.com, then click on the *More* tab, and the *Even More* menu item. Click to download and install Google Chrome.
> For Apple Safari - Visit http://www.apple.com/safari/download/.

Your computing experience needs to be a rewarding one. It will help significantly that you follow tips and learn new tricks to help boost the experience.

Conclusion

It has been a great pleasure for me to present what I know about the technologies around us, to research further on what I did not know best, and to deliver the results in this book to others who may need the information most. This is my first book, and I'm so proud and happy that I endeavored this project. I think I will do it again.

It is my fervent hope that the book has provided you, the reader, with information you wished you would acquire. For the classroom student, especially one in an area that does not have easy access to the latest technologies as they evolve, I hope this book has helped you to catch up on information together with those who reside in well-developed countries. Another such way of getting information, of course, is using the Internet if and when you can. The Internet is now a reservoir of a wealth of information on anything you want to know. In many ways, the Internet can be good, and in other ways, it can be bad. But you should continue to explore the Internet and watch out for its many pros and cons.

Also for the student, office worker, or merely a computer enthusiast, this book should've provided you with some exposure to the history of the computer we use today. In addition, you've been provided with tips, tricks, and shortcut methods of performing various tasks in your preferred computing environment for increased productivity.

References

1. The History of Computers in a Nutshell. Retrieved April 6, 2012 from http://sixrevisions.com/resources/the-history-of-computers-in-a-nutshell
2. TypesofComputerMemory.RetrievedMarch18,2012,fromhttp://www.computerknowledgeforyou.com/computercomponents/types_computer_memory.html.
3. University of Wisconsin, Intel & AMD Processors. Retrieved April 17, 2012, from https://kb.wisc.edu/showroom/page.php?id=4927.
4. Wikipedia, Color-coding scheme for connectors and ports. Retrieved May 7, 2012, from http://en.wikipedia.org/wiki/PC_System_Design_Guide.
5. Computer Hope, Microsoft Windows CE. Retrieved January 22, 2012, from http://www.computerhope.com/wince.htm.
6. Computer Hope, Microsoft Windows 3.1 and 3.11. Retrieved January 22, 2012, from http://www.computerhope.com/win3x.htm.
7. Computer Hope, Microsoft Windows 95. Retrieved January 22, 2012, from http://www.computerhope.com/win95.htm.
8. Computer Hope, Microsoft Windows 98. Retrieved January 20, 2012, from http://www.computerhope.com/win98.htm.
9. Computer Hope, Microsoft Windows ME. Retrieved January 22, 2012, from http://www.computerhope.com/winme.htm.
10. Computer Hope, Microsoft Windows NT. Retrieved January 22, 2012, from http://www.computerhope.com/winnt.htm.
11. Computer Hope, Microsoft Windows 2000. Retrieved January 22, 2012, from http://www.computerhope.com/win2000.htm.
12. Computer Hope, Microsoft Windows XP. Retrieved January 22, 2012, from http://www.computerhope.com/winxp.htm.
13. Computer Hope, Microsoft Windows Vista. Retrieved January 22, 2012, from http://www.computerhope.com/software/vista.htm.
14. Computer Hope, Microsoft Windows shortcut keys. Retrieved April 17, 2012, from http://www.computerhope.com/shortcut/windows.htm.

15. Microsoft Corporation, General keyboard shortcuts. Retrieved March 17, 2012, from http://windows.microsoft.com/en-US/Windows7/Keyboard-shortcuts.

16. Microsoft Corporation, Windows Logo keyboard shortcuts. Retrieved March, 17, 2012, from http://windows.microsoft.com/en-US/Windows7/Keyboard-shortcuts.

17. Microsoft Corporation, Windows Explorer keyboard shortcuts. Retrieved March, 17, 2012, from http://windows.microsoft.com/en-US/Windows7/Keyboard-shortcuts.

18. Microsoft Corporation, Microsoft keyboard shortcuts. Retrieved March 17, 2012, from http://windows.microsoft.com/en-US/windows-vista/Keyboard-shortcuts.

19. Microsoft Corporation, keyboard Shortcuts for Windows XP. Retrieved March 17, 2012, from http://www.microsoft.com/enable/products/KeyboardSearch_XP.aspx.

20. Microsoft Corporation, Using your keyboard. Retrieved February 16, 2012, from http://windows.microsoft.com/en-US/windows-vista/Using-your-keyboard.

21. Retrieved April 17, 2012, from http://windows.microsoft.com/en-US/windows7/Internet-Explorer-9-keyboard-shortcuts.

22. Microsoft Corporation, Windows Photo Gallery keyboard shortcuts. Retrieved April 17, 2012, from http://windows.microsoft.com/en-US/windows-vista/Keyboard-shortcuts.

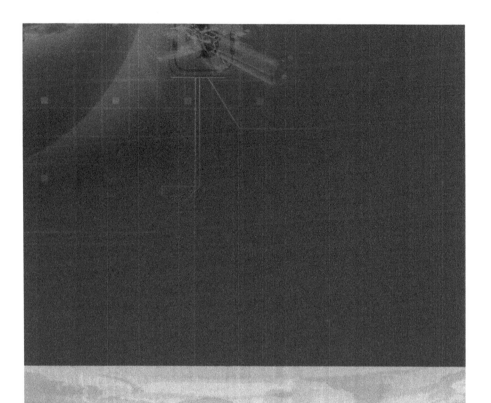

ISBN 978-1-4771-3871-7

90000

9 781477 138717

Xlibris

www.ingramcontent.com/pod-product-compliance
Lightning Source LLC
LaVergne TN
LVHW042341060326
832902LV00006B/312